W9-BMJ-857

TRADING WITH THE ENVIRONMENT

TRADING WITH THE ENVIRONMENT

Ecology, Economics, Institutions and Policy

Thomas Andersson, Carl Folke
and Stefan Nyström

EARTHSCAN
Earthscan Publications Ltd, London

First published in 1995 by
Earthscan Publications Limited
120 Pentonville Road, London N1 9JN

A catalogue record for this book is available from the British Library

ISBN: 1 85383 260 X paperback 1 85383 264 2 hardback

Typesetting and figures by PCS Mapping & DTP, Newcastle upon Tyne
Printed and bound in Great Britain by Biddles Ltd, Guildford and Kings Lynn

Earthscan Publications Limited is an editorially independent subsidiary of Kogan Page Limited and publishes in association with the International Institute for Environment and Development and the WWF-UK.

The Authors

Thomas Andersson is Associate Professor in Economics at the Stockholm School of Economics, and Head of the Structural Policy Secretariat at the Swedish Ministry of Industry and Commerce.

Carl Folke is Associate Professor in Systems Ecology at Stockholm University, and Deputy Director of the Beijer International Institute of Ecological Economics, The Royal Swedish Academy of Sciences.

Stefan Nyström is Director and coordinator of research in environmental economics at the Swedish Environmental Protection Agency. In 1990–94 he was responsible for trade and environment issues in the OECD and GATT at the Swedish Ministry of the Environment and Natural Resources. He is also vice chair of the Group for Economic and Environmental Policy Integration at the OECD.

Contents

CHAPTER 4

The New Playing Field – Towards Sustainable Development

List of Illustrations

FIGURES

BOXES

TABLES

Preface

Recent decades have seen a dramatic expansion of the human dimension on Earth. This expansion has placed stress on nature's life-support systems. There are few signs of reduction in environmental stress; on the contrary, most research suggests that the health of the environment will continue to deteriorate. Growing human populations and economies increase the pressure on life-support ecosystems, such as forests, coastal areas, agricultural land, rivers, lakes and oceans. Some people argue that humanity is far away from environmental constraints, others claim that global carrying capacity has already been exceeded. But, in any case, there are insufficient social mechanisms in place today to cope with these problems.

What part does trade play on this scene? Should there be firm restrictions on trade to protect the environment? Or would the environment benefit from free trade? Should an importing country only care about environmental impacts caused by domestic consumption, or does it also have responsibility for environmental degradation in foreign countries caused by production of the traded goods?

This book, which builds on a combination of different perspectives, addresses such issues from an integrated ecological economic approach. Such an integrated systems approach implies that the ecological preconditions provide the framework for economic development, and the actors in society ought to use all scarce resources, including environmental resources, in the most efficient way within this framework. Social norms and rules, the institutions that recognize the necessity to account for ecological realities, will encourage people to use nature's life support systems without degrading them.

In the book we discuss the conditions under which international trade could contribute to ecologically sustainable economic development, and propose certain changes to current regulations that would promote such a course. We do not claim to cover the whole picture in our analysis of this wide-ranging and complex area. Our hope is that this book will promote interest in crucial issues that will be in focus in both trade and environmental policy making far into the future.

Finally, it should be made clear that the views expressed in this book are the responsibility of the authors alone.

Thomas Andersson, Carl Folke and Stefan Nyström
Stockholm, Sweden
20 March 1995

Acknowledgements

Two months after the United Nations Conference on Environment and Development in 1992 in Rio de Janeiro, the Environmental Advisory Council to the Swedish Government organized a follow-up meeting on how to transform the challenges of the Rio Declaration and Agenda 21 into action.; 'Our task after Rio: a Swedish plan of action for the next century.'[1] The meeting addressed several topics, each discussed by a diversity of stakeholders from all over Sweden. One of the topics was trade and the environment. The work and discussions that the meeting generated illustrated the need for analysing the complex issue of trade and the environment. This inspired us to write this book, with the support from the Environmental Advisory Council, in particular Mrs Gunnel Hedman, at that time head of the Council. The first version of the book was published as a Swedish Government Report, in July 1993. A preliminary English version was distributed in a limited number in 1994.[2] This book represents a modified and updated version of the Swedish Government Report.

We are very grateful to Gunnel Hedman for her enthusiasm during this project. Gunnel Nycander has been another key person. She contributed to the Swedish report, produced most of the boxes and served as the technical editor. Eila Larsson, at the Environmental Advisory Council, has been a third key person. Paul Jannke, from Boston, USA, worked hard to improve the translation into English. Several persons provided constructive criticism on the manuscript. Astrid Auraldsson and Christina Leijonhufvud at the Beijer Institute assisted us during the work with the book. The project was originally supported by a grant from the Environmental Advisory Council. The Beijer International Institute of Ecological Economics and the Industrial Institute for Economic and Social Research (IUI) provided additional support. Carl Folke's work was partly financed through a grant from the Swedish Council for Forestry and Agricultural Research (SJFR), Thomas Andersson's by support from the Swedish Research Council for the Humanities and Social Sciences (HSFR).

1 'Vår uppgift efter Rio: Svensk handlingsplan inför 2000-talet' Report from the Environmental Advisory Council. Swedish Ministry of the Environment and Natural Resources. SOU 1992:104

2 'Handel och Miljö: mot en hållbar spelplan' Report from the Environmental Advisory Council. Swedish Ministry of the Environment and Natural Resources. SOU 1993:79. English version SOU 1994:76

Introduction

WHY TRADE AND THE ENVIRONMENT?

Towards the end of the 1980s, international interest in the links between trade and the environment grew significantly. A major factor contributing to this interest was a dispute between the United States and Mexico over American trade restrictions on imports of Mexican tuna. The United States believed Mexico was taking insufficient precautions to prevent the accidental catching of dolphins, and therefore placed an embargo on Mexican tuna in 1990. Mexico protested against this measure, claiming that the US-imposed sanctions were not compatible with either international law or the GATT (General Agreement on Tariffs and Trade) rules.

This particular incident was only the tip of the iceberg. International industry, as well as a growing body of public opinion concerned with the health of the environment, began demanding new measures against countries which were considered to behave in an environmentally irresponsible manner. Several EU (European Union) countries have, for example, discussed trade restrictions against imports of tropical timber. In Sweden, import restrictions regarding blue-finned tuna, a species threatened with extinction, have been debated. There are a variety of reasons why trade and the environment are becoming increasingly intertwined. A number of them will be presented in this chapter.

Different perceptions of the problem

Some environmentalists regard trade itself as the root of the environmental problem. This belief has led to a desire to impose far-reaching restrictions on the import of goods produced using environmentally harmful technologies. For example, the methods used in breeding or capturing certain animals have led to a demand for import restrictions.

Business interests in some countries advocate measures against the import of goods which are produced at a lower cost due to inferior environmental standards. Often, however, a tendency towards protectionism is the real reason behind this argument for environmental protection, for example measures taken to favour domestic industry in relation to foreign competitors. On the other hand, environmental measures affecting nationally produced goods in the domestic country are generally met with scepticism by industry, which protests that such measures will cause a decrease in employment and the companies' eventual relocation to another country due to the high cost of measures to protect the environment.

Trade experts are primarily concerned that the liberalization which has occurred during the last 40 years, largely as a result of the GATT agreements, will be undermined by environmental measures which will lead to a world-wide reduction in the flow of goods. Another aim of trade policy is a desire for 'harmonization' of regulations. This goal is less likely to be realized due to differences in environmental conditions, sensitivities and needs in differing countries.

In developing countries concern over trade restrictions has increased in the face of comprehensive demands for product information, especially in relation to packaging. A swarm of symbols such as swans, evergreens, falcons, angels, dandelions and green dots are appearing more and more on packages and products as a marketing prerequisite for markets in developed countries. Less developed countries regard this trend as a new form of protectionism, since they have difficulty both in obtaining information on such classification systems and in fulfilling the demands.

Trends within the environmental area

Manufactured goods have an increasing influence on the environment. Many of the environmental problems which were identified during the 1960s and 1970s, and above all emissions from large point source polluters, are at least partially on the way to being solved in industrialized countries. For example, Sweden reduced emissions from the thousand largest sources by 70 per cent during the 1970s and 1980s.[3] At the same time as point source emissions were being reduced, new environmental issues began appearing. The new problems are often connected to the array of new products constantly being developed and marketed.

In step with the increased consumption of goods, environmental problems caused by the utilization and disposal of consumer goods are growing. Many goods contain potential environmental risks, in the form of direct or delayed emission of harmful substances. There is, therefore, an increasing need to design and adapt products to the ecological cycle through recycling, re-utilization, and the efficient use of both energy and materials. In general, measures against large point source emissions only have an effect on the domestic manufacturing processes. Consequently, environmental policy measures enacted today have a more direct effect on trade with other countries. Measures regulating the environmental effects of consumer goods also influence foreign producers.

Secondly transboundary environmental problems are increasing. Attention has shifted from local environmental issues to regional and global problems. More diffuse and elusive challenges such as depletion of the ozone layer, the greenhouse effect and threats to biological diversity, are 'replacing' local envi-

3 'Hazardous waste' Swedish Ministry of the Environment and Natural Resources. Ds 1992:58

ronmental concerns previously symbolized by filthy sewers and stinking smokestacks.

Trends within trade

The volume of international trade is expanding: during this century, the volume of international trade has grown more quickly than international production. Diverse countries' economies are increasingly mutually interdependent. A country which trades with other countries becomes dependent on their trading partners' production systems and patterns of consumption. Such dependence results in the ability of countries to influence each other's economies. Trade is especially important for small, open countries.

The overall picture

The present tendency within trade policy and environmental policy is that measures in one will increasingly, and often inevitably, have a marked effect in the other.

INTERNATIONAL COOPERATION

Knowledge of the interconnectedness of trade and environmental policies has resulted in comprehensive work beginning in several international forums with the explicit goal of further exploring this relationship.

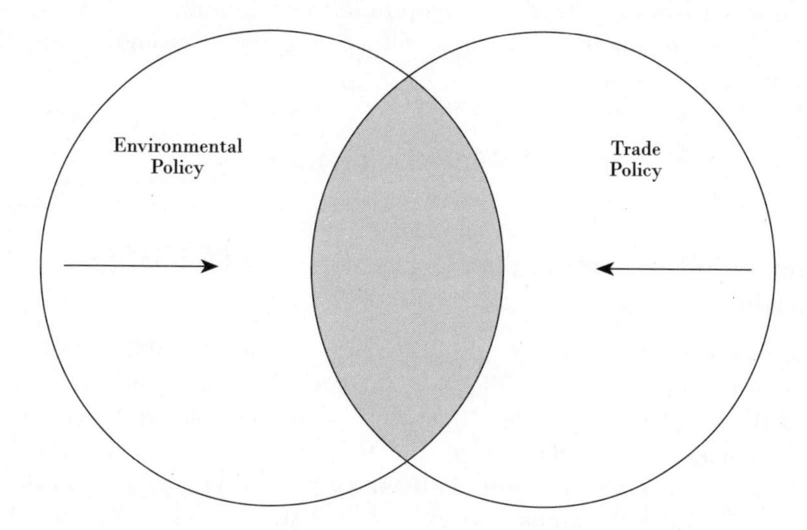

Figure 1 Trade policies and environmental policies increasingly overlap one another

Since the autumn of 1990, trade and environmental policy experts in the OECD (Organisation for Economic Cooperation and Development) have been working side by side to build mutual understanding. One of their goals is to identify potential problem areas and to create an information base which can be used during forthcoming GATT negotiations. The group has established specific guidelines for trade and environmental policy, with the aim of avoiding conflicts. In a communiqué from the OECD annual meeting between trade and finance ministers (June 1993), the group's goal was clearly stated: create an analytical foundation for the coming negotiations on trade regulations in other international forums.

Between 1991 and 1994, a technical group carried out work on trade and environmental issues within GATT. The group had a specific agenda[4] and, within the framework of its work, offered GATT member countries an opportunity to exchange views and present up-to-date analysis related to trade and environmental policy. In mid-1994, a specific environmental committee was created, in connection with the establishment of the World Trade Organization (WTO). The mandate of the committee was to open up negotiations concerning changes of trade rules. This implies that for the first time in history environmental issues have been given equal weight to other issues dealt with within the GATT/WTO; they have received an institutional platform.

UNCTAD (United Nations Conference on Trade and Development) has launched a comprehensive work-programme on trade and the environment, and UNEP has published several booklets and studies on the subject. There is a continually growing interest in the subject among non-government organizations and within the scientific community. Obviously, trade and the environment is already on the agenda and will most likely remain an important issue throughout the world.

OVERVIEW OF THE BOOK

Chapter 1: Nature's life support systems as the foundation for international trade

The importance of ecological systems as the basis for all economic activities is discussed in the first chapter. The role of the environment and the ecological services it provides are presented, as well as society's dependence on them. Environmental values from the perspective of economic theory are briefly given. In addition, economic growth, technological developments and the environment are discussed, and an ecological–economic synthesis is presented. This is followed by a description of international trade from a natural resource and environmental viewpoint. The significance of production and consumption pat-

4 1:Trade measures in international environmental agreements. 2: Transparency. 3: Packaging and marking.

terns, society's carrying capacity and ecological thresholds, transportation and various driving forces in relation to increased internationalization are some of the themes explored. The overall goal is a society that develops within the framework of functioning life-support systems, and that uses trade as a contributor towards such a development.

Chapter 2: Economic perspectives on trade and the environment

Economics is the theory of how scarce resources can be used to their greatest advantage. The environment can be included within the definition of resources; it is a resource base on which society and economic development rests, and which is deteriorated as degradation increases. To some extent, economics is about seeking the means to achieve certain goals, for example to increase human welfare. Therefore, one role of economics is to assist in finding suitable means to shape international trade so that it results in sustainable development. In this chapter, a short introduction to basic trade theory is presented. The chapter then examines whether all countries should have the same environmental protection regulations. Further, the arguments for introduction of trade barriers under various conditions (environmental effects due to consumption or production, and local or global environmental effects) are explored. Finally, the theoretical results are discussed in relation to the opportunities and risks posed by political reality.

Chapter 3: Trade regulations – the institutional framework and actual conditions

This chapter begins with an exploration of the concept of free trade, followed by an overview of the ground rules for international trade outlined by GATT, and also by the EU and NAFTA (North American Free Trade Agreement). In particular, the available opportunities for allowing environmental protection measures (which will affect trade) to take precedence over more general trade regulations, and what is most likely to happen in case of conflict, are examined and described. The growth of free trade in relation to environmental policies is then discussed in general terms. Finally, with this analysis as a departure point, we discuss how trade restrictions should be developed in order to realize the measures needed to achieve ecologically sustainable economic development while at the same time retaining stable and respected regulations for international trade.

Chapter 4: The new playing field – towards sustainable development

The final chapter summarizes the prerequisites for sustainable development. A number of ground rules are presented which must be observed for trade to contribute to sustainable development. The need is stressed for the market to receive signals regarding actual environmental values and costs. These values

and costs are often absent today. The significance of improved international cooperation is emphasized, as well as the mutual interest of trade policy and environmental policy in formulating regulations of trade which will hasten progress towards ecologically sustainable economic development. Finally, we present 16 conclusions.

Chapter 1

Nature's Life Support Systems as the Foundation for International Trade

Should society prioritize measures against environmental damage? Should we not put our own economic house in order before we can afford to care about nature and the environment? These questions are still quite common and reflect how the development of society has mentally distanced us from our dependence on healthy and functioning ecosystems.

The dependence of our society and its welfare on nature's life support systems will be described in the first section. This is followed by a discussion of economic growth and the environment and their relation to production, consumption and environmental technology. The next section explores the economic efficiency of society, the economic valuation of the environment, and the role of functioning ecosystems as increasingly scarce resources. The section ends with an ecological–economic synthesis for sustainable development. The last section offers an analysis of the role of trade in relation to ecological constraints, vulnerability, transportation and technical development.

LIFE SUPPORT ECOSYSTEMS – A PREREQUISITE FOR WELFARE

Throughout the last century we have increasingly lived with the belief that we are above nature; that we have the ability to decide the value, if any, of nature. In cities and industrial societies there are few visible signs of our day-to-day dependency on nature. When problems with natural resources and the environment occur, we turn towards our ingenuity, wealth of inventions and technological developments. Nature and the environment are viewed as something outside society, which can be replaced by technology.

Fortunately, this view of the world is changing, albeit quite slowly in many areas. We say fortunately, because life support systems are the basis of, and a prerequisite for, our welfare. Human beings are dependent on and a part of the ecological cycle, whether we recognize it or not. Because of our interdependence with ecological systems[5] it is not only the environment which is saved when society strives towards sustainable development, but also present and future generations of human beings.

There are signs which indicate that the world view which regards humanity as above and independent of nature, is changing. People's understanding is growing, whether they are from the North, South, East or West; we are realizing that we are a sub-system of the global ecosystem, the biosphere.[6] Survival of the human sub-system is dependent on the functioning of the overall system. Human beings need energy, natural resources and sustenance from the ecosystem on a daily basis in order for society and the economy to function.[7] Without natural resources it is not possible for machines, cars and other technologies to be built.[8]

We require food, air and water to live. Nature cannot be entirely replaced; humans cannot overcome dependency on the life-supporting environment. 'Natural capital'[9] will always be needed, and human beings will always be dependent on it. Therefore, we must conserve, rather than use up, this capital.[10]

Life-supporting ecosystems and ecological services

The life-supporting environment is that part of the earth which supplies the biophysical preconditions we need in order to survive: water, air, food, other types of energy and mineral nutrients. Earth's many species are closely intertwined with their respective environments in an eternal cycle of time and space which also contributes to providing the preconditions essential for life itself, including human life. This constitutes nature's support for society and its economy. It is sustained by the ecosystems; species interacting and evolving with their environment.

5 An ecological system, or ecosystem, consists of populations and communities of plants, animals, and microorganisms dynamically interacting with their physical and chemical environment, while simultaneously interacting with adjacent ecosystems and with the atmosphere.

6 Costanza, R (ed) (1991) *Ecological Economics: The Science and Management of Sustainability* Columbia University Press; Daly, HE (1987) 'The Economic Growth Debate: What some Economists have learned but many have not' *Journal of Environmental Economics and Management* 14:323–36

7 Ehrlich, P (1989) 'The Limits to Substitution: Meta-Resource Depletion and a New Economic-Ecological Paradigm' *Ecological Economics* 1:9–16; Folke, C (1991) 'Socio-Economic Dependence on the Life-Supporting Environment' In: Folke, C and Kåberger, T (eds.) *Linking the Natural Environment and the Economy: Essays from the Eco-Eco Group* Kluwer Academic Publishers, Dordrecht; de Groot, R (1992) *Functions of Nature. Evaluations of nature in environmental planning, management and decision making* Wolters-Noordhoff, Amsterdam

8 Hall, CAS, Cleveland, CJ and Kaufmann, R (1986) *Energy and Resource Quality: The Ecology of the Economic Process* John Wiley and Sons, New York

9 Natural capital consists of ecological services, renewable and non-renewable resources. The first two are produced and sustained by ecosystems and the last derived from them.

10 Costanza, R and Daly, HE (1992) 'Natural capital and sustainable development' *Conservation Biology* 6:37–46

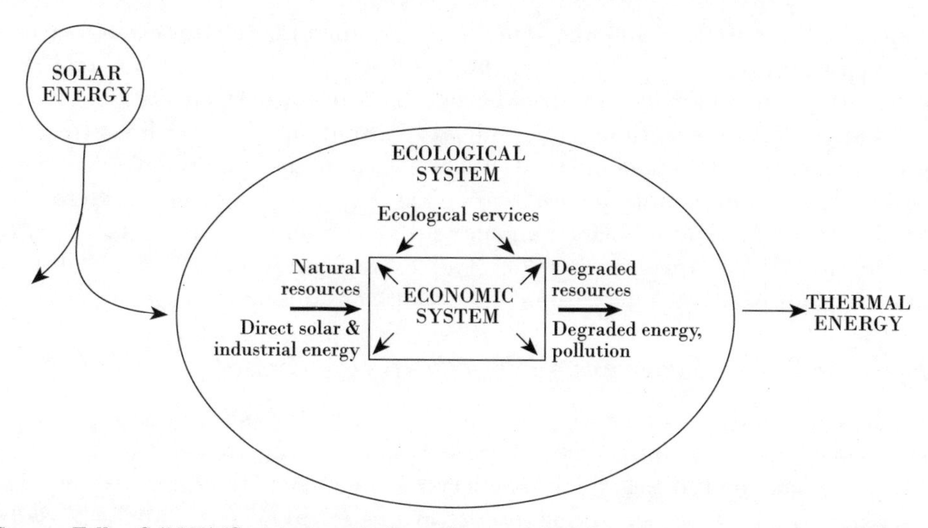

Source: Folke, C (1991) 'Socio-economic dependence on the life-supporting environment' In Folke, C and Kåberger, T (eds.) *Linking the Natural Environment and the Economy* Kluwer Academic Publishers, Dordrecht. Modified from Hall, CAS, Cleveland, CJ and Kaufmann, R (1986) *Energy and Resource Quality* Wiley-Interscience, New York; and Daly, HE (1977) *Steady-State Economics* Freeman, San Francisco.

Figure 2 Society and its economy is a sub-section of the ecosphere and needs energy, natural resources and ecological services. Ecosystems are factors of production supplying the human economy with ecological goods and services. Being parts of ecosystems, species sustain those crucial services.

Ecosystems produce renewable resources such as fish, crops, trees, drinking water and so on. They provide society with a variety of ecological services, for example the circulation of nutrients essential to agriculture, forestry and fishing. Ecosystems create fertile fields, pollinate crops, produce nourishment from the sea, maintain genetic diversity and the quality of the atmosphere, and provide pure air to breathe and water to drink. The ecosystems are 'in charge of' the crucial water cycle, and transform and also utilize part of the waste produced by society. Furthermore, all the functions and processes within ecosystems occur in a way which creates the basis for humankind's recreation in and enjoyment of the environment.

The dynamic structure of ecosystems forms the basis for nature's life support to society. Without functioning ecosystems, neither renewable resources nor ecological services would be generated. If soil acidification negatively affects the basic compounds required for flora and fauna, the ecosystem's ability to renew itself, and generate renewable resources and ecological services, is dramatically reduced. The production of ecological goods and services is, in other words, dependent upon a functioning and healthy ecosystem.

The importance of ecosystems to human welfare is rarely included in financial or development discussions. Only a negligible proportion of the renewable resources and ecological services is appreciated and an even lower proportion is considered to have an economic value.[11] Environmental debates often focus on the fish that are being poisoned or the trees that are suffering from acid rain rather than the diminished ability of ecosystems to support society.

This situation is serious, since society cannot function without healthy ecosystems. Despite our sophisticated societies, humans remain biological creatures who are undeniably part of ecological cycles. Life support systems are needed for a functioning society, and cannot be completely exchanged for capital or machines. Nature's support and ecological services are absolutely necessary for positive and sustainable development of human society.

Ecological boundaries and society's carrying capacity

Ecological (bio-physical) carrying capacity marks the maximum population size which can be sustained by nature within a specific area and under given technological capabilities. Societal carrying capacity is the maximal population size which can be sustained under various social systems, and especially the associated patterns of resource consumption.

In order to be able to approach sustainable development, societal carrying capacity must remain within the ecological carrying capacity.[12] This does not imply that ecological carrying capacity is static or absolute; it is dynamic. However, a society which degrades the ecosystem moves closer to the ecological thresholds and therefore reduces the potential for society's economic development.

Carrying capacity is not a static concept since ecological as well as human systems are constantly evolving. It is not possible to prevent them from changing. It is, however, possible to harm them. If ecosystem damage leads to irrevocable changes in nature's ability to sustain our society, then the choices available to our generation as well as future generations have been distinctly limited. The challenge lies in improving our management of the system's resilience; the ability to renew itself after unforeseen disturbances, to be able to absorb them, and to adapt itself to a changing situation.[13] The goal should be to retain flexibility and the freedom of choice to choose new paths of development in both the ecosystem and society.

How society uses resources and the limits of expansion

Production and consumption are often handled quite inefficiently today. Natural resources are pumped into one end of society and goods are produced. The goods are utilized and sooner or later discarded at the other end as waste.

This process is characterized by resources being collected from vast ecosys-

11 Gren, I-M, Folke, C, Turner R and Bateman, I (1994) 'Primary and secondary values of wetland ecosystems' *Environmental and Resource Economics* 4:55–74

12 Daily, GC and Ehrlich, PR (1992) 'Population, sustainability and Earth's carrying capacity' *BioScience* 42:761–71

13 Holling, CS (1986) 'The Resilience of Terrestrial Ecosystems: Local Surprise and Global Change' In Clark, WC and Munn, RE (eds.) *Sustainable Development of the Biosphere* Cambridge University Press, Cambridge, pp.292–317

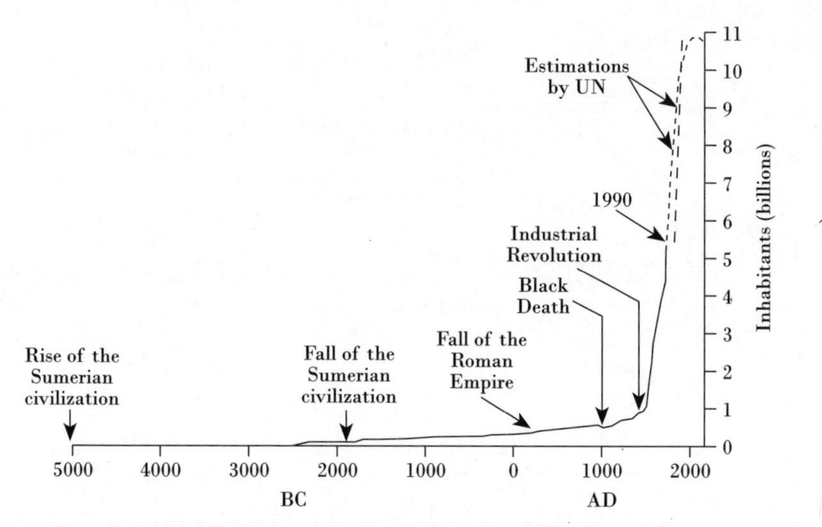

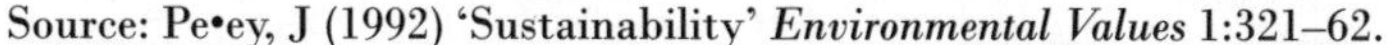

Source: Pe•ey, J (1992) 'Sustainability' *Environmental Values* 1:321–62.

Figure 3 The population explosion.

tem areas, often outside the country in question and concentrated within cities and industries. The concentration of resources in a limited area creates a situation where the surrounding ecosystems are not able efficiently to process the resulting pollution and waste products. Combined with the use of fossil fuel to run the infrastructure for industry and urbanization, this has led to a plethora of environmental problems.

The availability of inexpensive industrial energy, especially fossil fuels, has created this throughput-based, often wasteful, production system which lacks any connections to its life support ecosystems.[14] The way in which natural resources and fossil fuel have been used, rather than the fact that they are used at all, has caused severe environmental problems. History shows that more and more industrial energy, natural resources and ecosystems are needed in order to extract natural resources of increasingly lower quality. This has led to a deterioration of the natural environment. Often, even more industrial energy and natural resources are required to balance out the process.[15]

Developments in intensive agriculture, fishing and the timber industry are examples of this phenomenon. Soil erosion is compensated for in a variety of ways, such as the use of artificial fertilizer which is produced using fossil fuel. Decreasing fish populations are compensated for by the use of larger boats which cover larger areas of the ocean, and which are powered and produced by fossil fuels. In many cases fish populations are heavily exploited, which results in consequences for other economic sectors. Acidification has made it necessary

14 Odum, HT (1971) *Environment, Power and Society* John Wiley and Sons, New York
15 Bormann, FH (1976) 'An inseparable linkage: conservation of natural ecosystems and conservation of fossil energy' *BioScience* 26:759

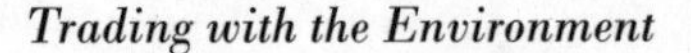

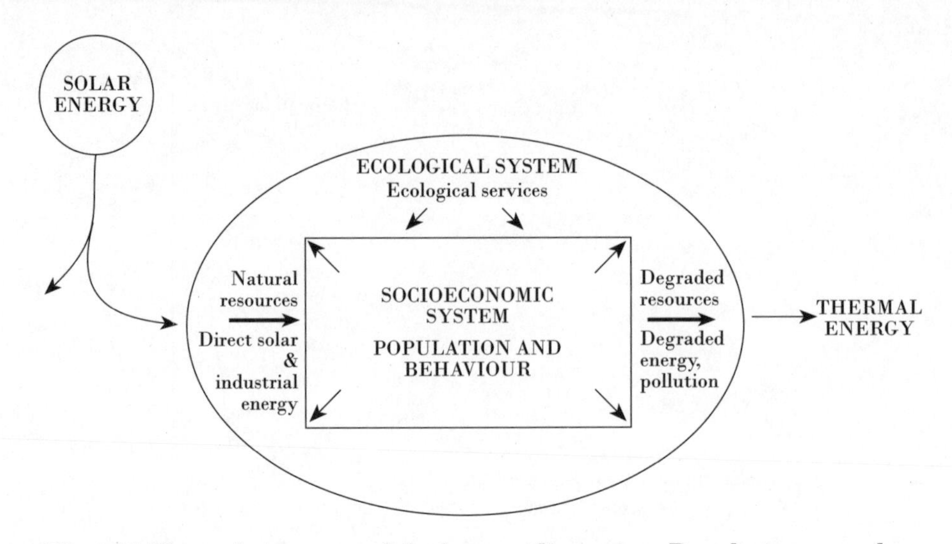

Figure 4 The scale increase of the human dimension. Population growth, expansion of human activities, and above all how they relate to the ecological resource base, have led to a situation where nature's ability to provide society with natural resources and ecological services increasingly restricts the development of society.

to 'doctor' the forests with products manufactured using fossil fuel. Therefore technical advancement is in itself no guarantee for a more intelligent and effective utilization of natural resources. However, given the right signals, technical development can be the key needed to turn society in the right direction.

At an earlier point in time, expansion of national economies was not limited, since their size was small in comparison to the global ecosystem's size and buffer capacity. Local environmental problems arose, but were often solved by techniques which simply moved them to other areas or other countries (for example higher smokestacks, extended pipe systems and transportation of waste). This method delays the solving of the problem at hand and leaves it in the lap of future generations (such as the collecting of sludge from sewage treatment plants, eutrophication and acidification).[16] Technological developments which freed economic developments from local ecological limits dominated. Scope for developing such technologies was an important limiting factor for national economies.

In the 1990s, the state of the ecosystems is increasingly important for societal development, and well-functioning life support systems become an even more limiting factor. This is due not only to the population explosion (Figure 3), but also to the fact that use of the earth's resources have increased enormously without any active method of revitalization or recycling.[17]

16 Thelander, J and Lundgren, L (1989) 'Nedräkning pågår. Hur upptäcks miljöproblem? Vad händer sedan?' Swedish Environmental Protection Agency

17 See, for example. the annual reports from World Resources Institute (*World Resources 92/93*) and Worldwatch Institute (*State of the World 1993*).

The population explosion, the increase in the scale of human activities and, above all, how these two factors have grown in relation to the ecological resource-base, have led to natural capital being increasingly regarded as a scarce resource for society's continued development (Figure 4). The longer it takes for us to recognize this, the more difficult it becomes to turn the whole process around, greatly increasing the risk for catastrophic results. We have begun to reach the limits of how much of our activities the global ecosystem can handle, which is reflected in such things as changing climate, depletion of the ozone layer and other serious environmental threats. Human beings alone use 40 per cent of the land ecosystem's life-supporting ability today[18] and the world population is expected to double in approximately thirty years.[19] This does put the scale of the human presence on the planet in perspective. Bearing these factors in mind, it is obvious that we must take good care of our own life support systems in order to sustain our existence; we should use them but not abuse them.

GROWTH AND THE ENVIRONMENT

A strongly rooted view in Western culture is that economic growth is a goal in itself, without regard to how growth is achieved. The presumption is often made that economic growth and liberalization of international trade are good for the environment. An example of this is Principal 12 of the Rio Declaration: 'States should cooperate to promote a supportive and open international economic system that would lead to economic growth and sustainable development in all countries, to better address the problems of environmental degradation.'

The general proposition that economic growth is good for the environment has been justified by evidence that there exists an empirical relation between *per capita* income and some measures of environmental quality. It has been shown that as income goes up there is increasing environmental degradation up to a point, after which environmental quality improves. As this is consistent with the notion that as their income rises people will spend proportionately more on environmental quality, economists have conjectured that it applies to environmental quality in general.[20] But, so far, that relationship has been seen only for a selected set of pollutants.[21] The results imply neither that economic growth is sufficient to induce environmental improvement, nor that the environmental effects of growth may be ignored, nor, indeed, that nature's life support systems are capable of sustaining indefinite economic growth. The

18 Vitousek, PM, Ehrlich, PR, Ehrlich, AH and Matson, PA (1986) 'Human appropriation of the products of photosynthesis' *BioScience* 36:368–73

19 Ambio (1992), Special Issue: Population, Natural Resources and Development 21:70–122

20 Beckerman, W (1992) 'Economic growth and the environment: whose growth? whose environment?' *World Development* 20, 481–96

21 Shafik, N and Bandyopadhay, S (1992) 'Economic growth and environmental quality: time series and cross country evidence' Background Paper for the World Development Report, World Bank, Washington

Box 1
ENVIRONMENTAL TECHNOLOGY: A GROWTH SECTOR FOR THE 1990S

The environmental industry is growing quickly, and today the market has reached 200 billion US dollars. There are no clear parameters for this area as it includes a wide range of various purification techniques, products and consulting services which are registered in many different industries. The market for water conservation is the oldest, and Europe (and specifically Germany) leads in that arena, while the USA dominates the waste disposal area and Japan leads air conservation.

Table 1 Sectors of the environmental technology industry: relative importance and expected growth

	Relative distribution within OECD	*Expected annual growth, %*
Pollution-abatement	76	5.0
Equipment and services		
of which: water conservation	29	4.0
air	15	4.4
waste & recycling	21	6.4
miscellaneous	11	5.1
Consulting services and clean technology	24	7.4
Total	100	5.5

Source: OECD (1992) *The OECD Environment Industry: Situation, Prospects and Government Policies* OECD/GD (92) 1, Tables 1 and 6.

The whole environmental sector is expected to grow by approximately 5.5 per cent a year, and to reach 300 billion US dollars by the year 2000. This can be compared to the chemical industry which should reach 500 billion US dollars. Air and water conservation are the two oldest and best established markets. The sectors which are expected to grow the most quickly, however, are waste disposal, recycling techniques, environmental consulting services and clean technological developments. These areas are growing while at the same time 'end-of-pipe solutions' are declining and conservation of the environment is integrated into the entire production process. Sweden is well-placed in two sectors: Alfa Laval is one of the three leading companies in water technology, as is Fläkt for air purification.

Germany has been heavily involved from an early stage and today holds a third of the European environmental technology market. Their market is expected to grow at a slightly slower rate, 4 per cent. The markets are growing most rapidly in Canada (7.9 per cent), England (6.3 per cent), Japan (6.7 per cent), and in Portugal, Greece and Spain (7.4 per cent to 8.3 per cent). Sweden is further down on the list with 3.7 per cent, with only Switzerland, Finland and Denmark placed lower.

Table 2 The market for environmental technology: volume 1990 and growth

	North America	*Europe*	*Japan*	*Sweden*	*OECD*	*All countries*
Total market, billion US dollars	84	54	24	1.5	164	200
Expected annual growth %	5.4	4.9	6.7	3.7	5.5	5.5

Source: OECD (1992) *The OECD Environment Industry: Situation, Prospects and Government Policies* OECD/GD (92) 1, Table 7.

The market is quite dependent on government policies, since environmental policies create new markets. The largest and most technologically advanced companies have developed in countries with strict environmental legislation. Other compelling forces for companies to acquire environmental technology are the desire for a good reputation – not just with the public, but also with employees, customers, employers and investors. Growing understanding of both the financial and general value of environmental conservation occasionally causes companies to precede legislation in this area. Having an environmentally-friendly product is recognized as an increasingly important competitive strategy.

A total of 1.7 million people worldwide work in environmental technology companies including 800,000 in the USA, 250,000 in Germany and 200,000 in Japan. It is a highly international business area, with a large proportion of international trade. Germany exports 40 per cent of its production, other net exporters in Europe are England, Sweden, France and Holland.

The environmental technology sector is strategically important, partly because it is growing and partly because it is becoming a prerequisite for growth and competition in other sectors. Apart from strict domestic legislation, research and development are important factors needed to develop competitive environmental technology companies. Today 2 billion US dollars, or 2 per cent of the governments' available research funds within OECD, are invested in environmental research. Sweden is above the average with 2.5 per cent, but behind Denmark (3 per cent), Holland (3.8 per cent), and Germany (3.4 per cent).

Source: OECD (1992) *The OECD Environment Industry: Situation, Prospects and Government Policies* OECD/GD (92) 1.

have to be reduced by about 75 per cent before a gradual loss of functional diversity would flip the system to a different pattern of behaviour. Radar images of flights of migratory birds across the Gulf of Mexico over approximately a 20 year period have revealed that the frequency of trans-Gulf flights have in fact declined by almost 50 per cent, approaching the range of uncertainty in the simulation estimate above. Hence, Canadian Boreal forests and the economic activities dependent upon their functioning, are, in addition to local landscape fragmentation, threatened by the human population growth and land-use pressures in neo-tropical countries and along the migration paths of insectivorous birds.[34]

Other examples of such ripple effects of local damage to the environment include:

- When mangrove forests on the coasts of South America and Asia are chopped down, this results in decreased fish reproduction, and thereby fewer fishing catches for other countries exploiting the fish stocks.
- An increase in the density of cattle in Holland causes downfall of nitrogenous compounds in the surrounding countries.
- The destruction of wetlands leads to eutrophication of coastal zones and coastal waters.
- Landscape modifications have caused outbreaks of new diseases by shifting the pattern and abundance of insects.
- Coastal pollutions have restructured plankton communities causing toxic plankton blooms, as well as outbreaks and regional spreading of diseases including cholera.[35]

Many of the ecosystem modifications and the environmental effects that they cause, which today are regarded as only local or national, are in fact already of regional and ultimately global concern. The web of connections linking one ecosystem to the next is intensifying on all scales in both space and time. Local human influences on air, land and oceans slowly accumulate to trigger abrupt changes when ecological thresholds are reached, directly affecting ecological services and the vitality of human societies elsewhere. Local environmental problems often have their cause half a world away, as illustrated by the above examples. Everyone is now in everyone else's backyard.[36]

Local environmental damage can also influence other countries through negative effects on the social structure. Examples of this are deforestation,

34 Holling, CS (1994) 'An ecologist's view of the Malthusian conflict' In Lindahl-Kiessling, K and Landberg, H (eds.) *Population, Economic Development and the Environment* Oxford University Press, pp. 79–103
35 Almendares, J, Sierra, M, Andersson, PM and Epstein, PR (1993) 'Critical regions, a profile of Honduras' *The Lancet* 342:1400–1402; Epstein, PR, Ford, TE and Colwell, RR (1993) 'Cholera-algae connections' *The Lancet* 342:14–17
36 Folke, C, Holling, CS and Perrings, C (1995) 'Biological diversity, ecosystems, and the human scale' *Ecological Applications*, in press

over-exploitation of farmland and specialization in export crops.[37] These can compromise the resource base for large sections of the population, cause an increase in the population and even cause 'environmental refugees' to flee their homeland. Discovering ways of developing society which promote ecological security is therefore of great importance.[38]

Local environmental damage can also lead to changes in welfare in foreign countries, regardless of whether or not their ecosystems are affected. The extinction of a certain species may, for example, lead people in other countries to feel such sadness that their well-being is damaged by the extinction. The costs of preserving species in national parks in developing countries has shown itself to be high on a local level, lower on a regional or national level and lower still on a global level. When it comes to the revenue from or value of saving a species in a national park, the situation is the opposite.[39] Purely aesthetic aspects of our relationship to the ecological system, biological diversity and other human beings' quality of life often mean that it may be justified to treat local problems as global.[40]

In conclusion, many more environmental effects are transboundary and therefore international than those receiving attention today. This is a consequence of the increase in scale of the human population and its economic activities. The larger this dimension grows in relation to the the capacity of ecosystems to sustain it, the closer to critical thresholds we will get. In order to avoid challenging such thresholds, the effects of local environmental damage should be regarded and managed as being transboundary.

ECONOMY AND ECOSYSTEMS

An efficient economy implies that scarce resources are managed so as to achieve maximum benefit to society. According to economic theory, in a market with perfect competition, prices of goods and services reflect society's actual costs and benefits. In practice, there are few perfect markets. When different factors restrict or distort competition, market prices need not convey such information, and markets cannot properly reflect society's actual costs and benefits. Lack of information is also significant, markets cannot properly handle lack of information and risks. In addition, vital resources fall outside the market when there are no well-defined property rights. There can be no market price if 'sellers' do not possess full ownership and cannot defend their rights as owners. But who owns the climate? How much does a new ozone layer cost?

37 Khogali, MM (1991) 'Famine, desertification and vulnerable populations: the case of Umm Ruwaba District, Kordofan region, Sudan' *Ambio* 20:204–206
38 *Ambio* (1991) Special Issue: Environmental Security 20:168–206
39 Wells, M (1992) 'Biodiversity conservation, affluence and poverty: mismatched costs and benefits and efforts to remedy them' *Ambio* 21:237–43
40 Molander, P (1992) 'Frihandeln ett hot mot miljöpolitiken – eller tvärtom?' Swedish Ministry of Finance, Expertgruppen för studier i offentlig ekonomi, 1992:12, p.31.

Market prices therefore seldom reflect all the changes to society's welfare when goods or services are used. This results in external effects, or costs and benefits which are not accounted for in market prices. That they are not accounted for means that the individual or company responsible for a change is not confronted with the costs and benefits of their actions, even though their actions have resulted in a change in the welfare of others. An example of an external effect is the case of a factory which discharges toxic waste, killing fish in an adjacent lake without the factory owner having to compensate for the losses of the people who fish there. In this way, there will be a difference between private and social values. The private value is the change in welfare which the consumption of goods or services offers the individual, and the social value reflects the total changes in the welfare of all individuals, including both present and future generations.

The difference between private and social economic values is very important for the environment and life support systems. Environmental utilities rarely carry a price tag, whether directly at market level or indirectly through various policies.

Internalization of external effects

External effects can result in substantial costs for society. One way to reduce environmental problems from the viewpoint of economics is to internalize the external effects; to correct for the existence of externalities so that costs and benefits are debited to the agents who incur them. By internalizing environmental costs we mean all measures, direct and indirect, which influence the costs and benefits of the actors in society so that the difference between private and social value decreases, and ideally disappears.

Internalization means that the costs of environmental damage are included in the price of those goods and services which cause the damage. For example, the cost of cancer-related illnesses and harmful effects on plants and animals would be included in the price of products which cause depletion of the ozone layer. An increased internalization of environmental costs would change production into energy and resource-effective processes and products would be better suited to basic ecological conditions. New types of investment would be generated and the prerequisites for competition would change.

An indication of the degree of internalization of environmental costs throughout the world is given in the 1993 World Watch Institute report. The Institute annually presents information on the global state of the environment. A complete internalization of environmental costs would, in principle, mean that the boundaries of the long-term carrying capacity of the natural environment are not crossed in any single case.

Unfortunately, most signs point to an opposing trend: continued externalization. Global forest acreage is decreasing. According to the World Watch Institute, destruction of tropical rain forests is continuing, desertification is increasing, and one-third of the world's agricultural area suffers serious erosion problems. Furthermore, the number of plant and animal species is quickly being decimated (approximately three-quarters of the world's 9000 known species of birds are at risk of either depopulation or extinction). The concentration of 'greenhouse' gases in the atmosphere is increasing annually and virtually every study of the ozone layer shows that it is being depleted at an ever-increasing rate. Experience has shown that the market is not able to handle the internalizion of environmental costs on its own. Society therefore has a vital role to play. With the right signals from society, it would be profitable for companies and other key players to develop ecocyclical, resource-effective technology that works in cooperation with the environment.

Environmental costs can be internalized directly:

- via laws and regulation of products and production technology;
- via markets for emission rights, environmental charges and taxes;
- by controlling resource use and waste and pollution through better defined property rights and other institutional reforms.

It is more difficult to get an overview of indirect measures because of the complex nature of social structure and dynamics. Indirect measures can be created through changes in

- macro-economic policy;
- legislation;
- consumer behaviour;
- changes in social values.

In this context, the establishment of institutional frameworks, national and international, becomes extremely important. How the groundrules fot individuals and markets are shaped plays a decisive role in the internalization of environmental costs. The formulation of these rules relates to society's view of its relation to the natural environment. These aspects are discussed later in this chapter (see for example Figure 12). Various institutional frameworks are discussed in Chapter 3.

Valuation of changes in welfare

In order to determine the extent of externalities and thereby internalize them, a number of methods have been developed within environmental economics. The foundation for the valuation of costs and benefits is changes in human welfare, based on human preferences and expressed as people's direct or indirect willingness to pay.[41]

The methods and their application are limited because many environmental effects of human activities are not known at the time of valuation. In

41 Johansson, P-O (1987) *The Economic Theory and Measurement of Environmental Benefits* Cambridge University Press

general, the valuation reflects the perceptions of individuals of an externality and not the actual consequences on nature's life support systems that the externality causes. It is also difficult to take into account the values of future generations. Environmental effects tend to accumulate and first become visible many years after the original event, and the dimensions are often dependent on complex relationships to other external effects, as well as on the resilience of ecosystems. The amount of uncertainty creates a situation where various risks must be weighed against each another. Valuation is further hindered by the lack of real markets. Artificial markets either cannot or perhaps should not even be constructed in order to measure the willingness to pay.[42]

A complete valuation of external costs can not be achieved, of course. Nevertheless, valuation of environmental costs and benefits plays an essential role in bringing environmental concerns into economic decisions, both in relation to farming regulations and laws from society's viewpoint, and to companies' and consumers' actions. An example of this is salmon farming in cages in coastal areas, an industry which has expanded rapidly since the mid 1980s. Production is a linear throughput type which does not relate to the resource base. Salmon production has been shown to be ecologically unfeasible.[43] The process causes a variety of environmental effects, one of which is eutrophication of coastal waters. Manufacturers do not pay for the cost of the eutrophication to society. Calculations show that if the cost of eutrophication were to be internalized (based on society's willingness to pay for technology that prevents the eutrophication) then the cost of producing salmon would exceed the highest price the manufacturer received for salmon on the market. In this case, it is sufficient to internalize one of the many environmental effects in order to judge whether production is economically feasible or not.[44]

As mentioned earlier, ecological preconditions determine the framework for economic activity. Society can reformulate this in terms of absolute environmental goals, such as a ceiling on the amount of sulphur dioxide allowed in the air.[45] Bearing in mind the difficulty in valuing the environment in monetary terms, an important task is to find cost-effective ways of reaching these goals within the ecological framework.[46]

A functioning ecosystem: 'the new scarcity'

A number of ecological services are already scarce in the 1990s, even though

42 Bojö, J, Mäler, K-G and Unemo, L (1990) *Environment and Development: An Economic Approach* Kluwer Academic Publishers, Dordrecht
43 Folke, C and Kautsky, N (1992) 'Aquaculture with its environment: prospects for sustainability' *Ocean and Coastal Management* 17:5–24
44 Folke, C, Kautsky, N and Troell, M (1994) 'The costs of eutrophication from salmon farming: implications for policy' *Journal of Environmental Management* 40:173–82
45 Bishop, RC (1978) 'Endangered Species and Uncertainty: The Economics of a Safe Minimum Standard' *American Journal of Agricultural Economics* 60:10–18
46 Gren, I-M (1991) 'Costs for Nitrogen Source Reduction in a Eutrophicated Bay in Sweden'. In Folke, C and Kåberger, T (eds.) *Linking the Natural Environment and the Economy: Essays from the Eco-Eco Group* Kluwer Academic Publishers, Dordrecht

this is seldom visible in market prices. There are reasons to expect increased shortages. If the availability of clean air, clean drinking water and other ecological services was unrestricted, then they would not be scarce resources. There also would be no reason to include them in economic analysis. But nature's life support systems and ecological services are increasingly becoming scarce resources, in addition to the fact that they represent fundamental production factors for national economies.

Previously, a large proportion of life support and ecological services, for example the availability of water, fresh air, or tropical rain forests and wetlands, did not put constraints on economic development from a global perspective. The variety of life support systems and ecological services could buffer economic activity, and their supply was sizeable in relation to demand. Local limitations occurred nevertheless.

The demand for clean air, water and other ecological services has increased and has 'pushed' humanity closer to global ecological threshold effects. This is because both the earth's population and the consumption of resources have increased substantially. The way in which this has taken place has damaged the natural environment,[47] causing the availability of ecological support to diminish.

It is difficult to see the total consequences of many individual actions, and thereby to determine exactly when life support systems should be regarded as a scarce resource. We do know, however, that there is always true uncertainty and a lack of information when complex ecological and economic systems must be managed and that irreversible changes can develop when an ecosystem's resilience deteriorates.[48] The social costs for unforeseeable changes can be enormous and have an influence far beyond the national borders.

We do not know exactly where we are in relation to global ecological thresholds, but as we have already mentioned, it is no longer rational to believe that the capacity of ecosystems to support human society is still unrestricted, that the price is still zero. The reason for the difficulty in determining exactly when resources become scarce is that environmental effects are often delayed in both time and space. They accumulate and spread geographically and it takes time before the effect on nature's life support capacity becomes visible. It takes longer for an environmental problem to be identified and longer still before information leads to measures which send signals to society and business that there is a new scarcity – an ecological scarcity.[49]

Certain types of ecological services have become so damaged that by the time the damage is visible it is too late to start managing resources. From a global perspective, resources can be scarce because of different systems' interdependence, even though the same resources are not considered to be scarce locally and from the perspective of a certain sector. It is necessary to try to

47 Odum, WE (1982) 'Environmental degradation and the tyranny of small decisions' *BioScience* 32:728–29
48 Costanza, R, Waigner, L, Folke, C and Mäler, K-G (1993) 'Modeling Complex Ecological Economic Systems: Toward an Evolutionary, Dynamic Understanding of People and Nature'. *BioScience* 43:545–55
49 Barbier, EB, Burgess, JC and Folke, C (1994) *Paradise Lost? The Ecological Economics of Biodiversity* Earthscan, London

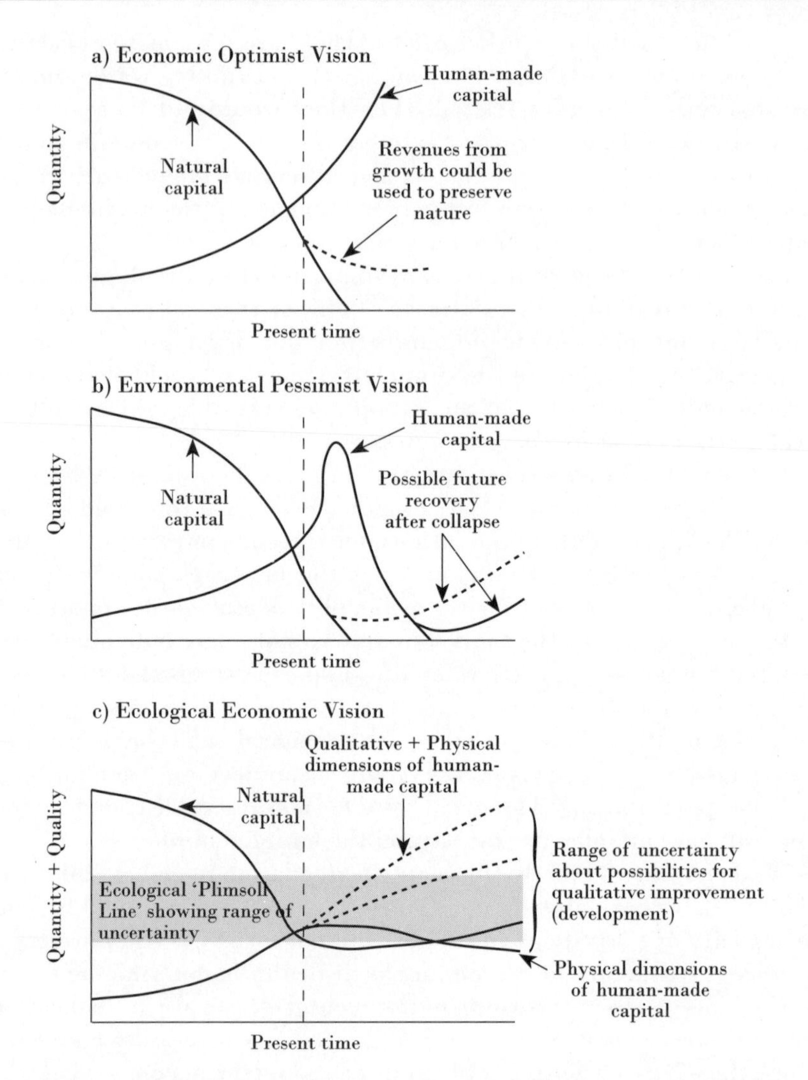

Source: Folke, C, Hammer, M, Costanza, R and Jansson, A-M (1994) 'Investing in Natural Capital: Why, What and How?' In Jansson, A-M, Hammer, M, Folke, C and Costanza, R (eds.) *Investing in Natural Capital: The Ecological Economics Approach to Sustainability* Island Press, Washington.

Figure 6 Three different perspectives of the socio-economy's relationship to nature. A) Growth and technical optimism. B) Environmental pessimism. C) The ecological–economic perspective.

foresee this and to avoid a situation in which there is serious ecological scarcity. 'Pro-active' rather than 're-active' management becomes necessary, with the goal to manage *all* of society's resources as efficiently as possible.

AN ECOLOGICAL–ECONOMIC SYNTHESIS

In Figure 6 we present a conceptual picture of the above discussion. The illustration shows the simplified relations between natural capital and manufactured capital, seen from three different viewpoints. In all three cases natural capital has decreased and manufactured capital increased. The dotted line indicates today's situation.

The first picture illustrates the attitude that society must first have economic growth in order to be able to afford to invest in the environment. This attitude is based on a world view where human beings believe that they are not constrained by nature, where nature and the environment are regarded as separate from the economy and where environmental problems are solved as they arise by developing new technology. According to this paradigm free trade encourages economic growth (Figure 6a).

In the second picture, human beings are seen as parasites whose society has destroyed nature. The limits of the earth's carrying capacity are considered to have been reached or already passed, and a decrease in all human activities is regarded as essential in order to avoid ecological collapse. Here, a move towards free trade is thought to increase environmental destruction (Figure 6b).

Both these viewpoints segregate society and the environment and do not examine the whole picture of the relationship between ecology and the economy. Such segregation does not lead to constructive solutions for the future and cannot guide society towards ecologically sustainable economic development. This book strives for an integrated systems view of the relationship between nature and human society, between ecology and economy.

The socio-economy represents a part of the overall ecological system. Life support systems determine the ultimate limits for society's expansion. The closer we get to these limits, or thresholds, the less room there is for economic development. The limits are not fixed, rather they are constantly changing, largely due to human behaviour towards the resource base. An important challenge is the task of turning societal development from a continued rapid expansion of material consumption towards a development which is resource-efficient and is connected to the functions and processes of the ecosystems. The ecosystems and ecological services that they provide must be appreciated as fundamental factors of production which, as increasingly scarce resources, need to be managed efficiently.

Growth and technical development by themselves cannot guarantee solutions. It is *how economic growth* (as a measurement of societal development) *occurs* and *in what direction technology develops* which determine if ecologically sustainable economic development will take place. Free trade in itself is no guarantee of an improved environment. Whether liberalizing trade will have a positive or negative effect on the environment depends to a great extent on the underlying institutional structure and economic development within which trade takes place.

This interplay can be illustrated by the following metaphor. If a large boat is awkwardly loaded, its ballast is uneven, and it cannot effectively transport its cargo of goods without the risk of sinking. We set up a market, a price system, where the price of loading the boat is higher or lower depending on how near or far to the waterline the boat is when it is being loaded. In this way, the market ensures that the boat is efficiently loaded. However, if we go on loading the boat with goods, it will sink. This is why the Plimsoll line (see Figure 6c) exists; it indicates that goods loaded over a specific mark will cause the boat to sink.[50]

The metaphor demonstrates that the market plays an important role in improving the management of natural resources and the environment and that the internalization of environmental values and costs is essential since it will encourage markets to work more efficiently than before, from a social perspective. But if the scale of activity constantly rises, then improving market efficiency is no guarantee of long-term sustainable development since in practice internalization cannot be complete, in part because of true uncertainty, market and policy distortions, and a lack of information. The importance of clear and basic norms and rules, and long-term signals from institutions in society increases as human society approaches ecological thresholds.

This growing importance is why we constantly emphasize that ecological constraints provide the framework and the preconditions for society and human welfare, and that society should strive to manage all resources in the most efficient way possible within this framework. Inefficient use of resources increases the risk of 'a sinking ship', the ecological thresholds come closer and freedom of choice and action continues to shrink. It is an important challenge to interpret and manage the signals we are getting from ecosystems, in order to avoid a collapse of jointly determined economic and ecological systems.

TRADE FROM A NATURAL RESOURCE AND ENVIRONMENTAL PERSPECTIVE

Trade magnifies economic development

Trade originates in markets, which in turn coordinate production and consumption of goods and services. Trade is therefore connected to the environmental effect which can occur when goods and services are produced and consumed. Trade between two countries influences the prices of goods and resources, income, production and consumption patterns, and so on in both countries. Since trade affects national economies and national economies affect the ecosystem, trade can be said to influence the environment indirectly. Whether that influence is positive or negative is difficult to determine in advance due to the

50 Daly, HE (1984) 'Alternative Strategies for Integrating Economics and Ecology'. In Jansson, A-M (ed.) *Integration of Economy and Ecology: An Outlook for the Eighties. The Wallenberg symposia* Askölaboratoriet, Stockholm University

complexity of the interlinked socio-economic and ecological systems.

If the institutional structure in a country ignores the fact that production of goods deteriorates life support systems, then trade with these goods will strengthen this trend. In other words, it will increase the environmental effects which occur when goods are produced and consumed. If institutions create ground rules which point production and consumption towards a sustainable development, then trade can add to this process.

> *None of the aspects of sustainable development is intrinsically linked to international trade. A failure to place a value on environmental resources would undermine sustainable development even in a completely closed economy. Trade is seen, rather, as a 'magnifier'. If the policies necessary for sustainable development are in place, trade promotes development that is sustainable.*
>
> *International Trade, 1990–91*, GATT 1992, p. 25.

A country like Sweden is highly dependent on imported goods in order to maintain its material prosperity. If production of imports destroys the environment then Sweden's demand adds to the degradation of ecosystems occurring in other countries. Globally, cotton raised for export accounts for 25 per cent of international consumption of insecticides; an instance where Swedish consumers indirectly contribute to an environmental problem through their purchase of cotton clothing.[51] The decisions of individual consumers seldom have the ability to influence production, but the collective impact of consumer demand does.

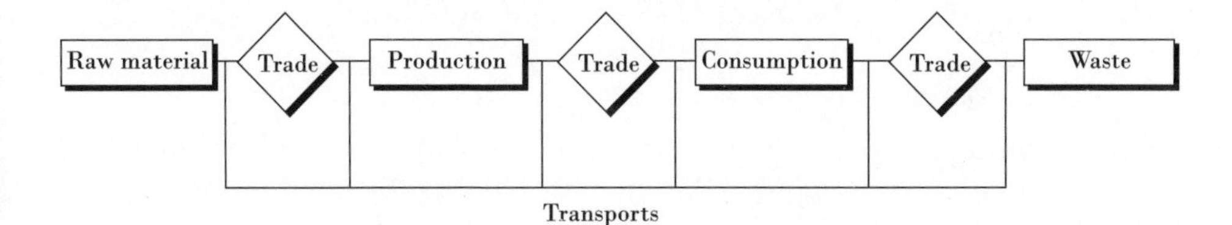

Figure 7 The role of trade in relation to the life-cycle of a product.

Figure 7 symbolizes a simple life cycle for one product, commonly referred to as 'cradle to the grave'. No single link in the life-cycle chain can be skipped; each link is necessary for production, trade, transportation and consumption to take place. Trade is an integrated part of the chain and therefore cannot be removed and regarded as a separate entity in a discussion of the different steps involved in environmental influence.

51 Madeley, J (1992) *Trade and the Poor. The impact of international trade on developing countries* Intermediate Technology Publications, London

Trade internationalizes the environmental problem

Environmental problems can be divided into such categories as, what happens during production, and what happens when goods are consumed or become waste. It is not easy to make this type of division in reality. At the intermediary stage, effects on the ecosystem from, for example, the petrochemical industry can be regarded as environmental effects from either oil consumption or plastic production. Both consumption and production effects can be local or transboundary. Using this schematic division, situations can be differentiated where the presence of trade contributes to the development of an international environmental problem (Figure 8).

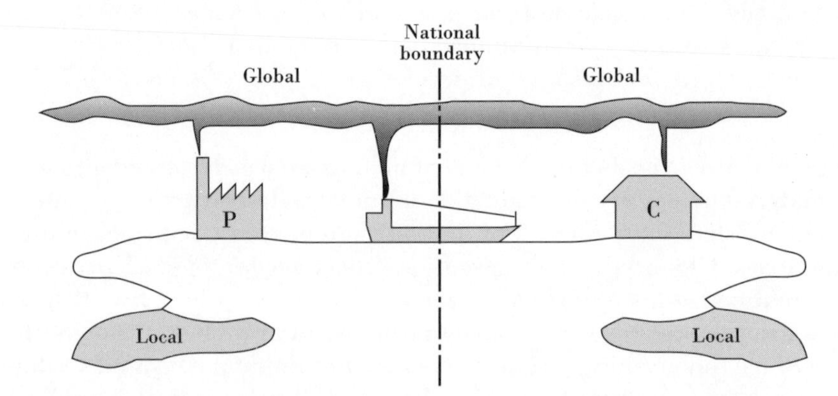

Figure 8 Simplified picture of production (P), consumption (C), local and transboundary environmental problems and how trade contributes to the internationalization of environmental problems.

Environmental problems become international, either when pollution and the consequences of ecosystem degradation cross national boundaries, or when the cause of the environmental effects – a product or a producer – crosses national boundaries. The connection between trade and the human impact on nature's life support systems can be organized in the following manner:

- Trade creates dependency on ecosystems of foreign countries.
- Trade has a direct effect on the environment, via transportation.
- Many imported and exported goods cause environmental effects when they are consumed. Waste is a special case.
- Environmental regulations and taxes affect production costs and competitiveness of goods which cause environmental effects during production. If environmental demands become too strong, there is a risk that production will cease in one country or simply move to another country with less stringent laws. Trade options can therefore cause producers to move out of the country or force policy makers to hinder

environmental legislation to avoid triggering this migration. This is the case for both local and global environmental effects but especially global effects, since the environmental cost of having little or no protective legislation is spread all over the world.

The first two points are discussed in this chapter. The other points are discussed in Chapter 2, where we will also examine the general question of how trade affects the environment and if and how trade barriers can be used to reach environmental objectives.

Trade dissolves local boundaries

Sustainable development requires society to remain within ecological boundaries. These boundaries exist on different temporal and spatial levels, from global (such as the greenhouse effect) to local (such as fish production in lakes).

Trade between different countries offers an opportunity for people to free themselves from local ecological constraints by importing resources (raw or processed materials) and ecological services from other countries. Access is thereby gained to information, capital, resources or products not available in the domestic country.

Even though trade dissolves national boundaries, it cannot do away with ecological boundaries. Ecological constraints can be moved but not eradicated and they are shared by many nations. Continued unsustainable production and consumption patterns threaten nature's capacity to sustain national economies and bring humanity closer to large-scale ecological thresholds, the risk being that our interdependent human societies will hit a ceiling simultaneously.

Hence, the majority of the world's economic communities are dependent not only on imports from other countries but also on the support of the ecosystem which produces those import goods. Trade means that people are indirectly dependent on their access to resources and ecosystem functions outside the boundaries of their own country. Trade moves ecological boundaries outside national boundaries, and large ecosystem areas, or shadow areas, are required to produce the imported goods and services.

Shadow areas

Estimating the spatial ecological support – or shadow area[52] – a city needs is one way to illustrate the importance of functioning ecosystems for human well-being. The shadow area is the area needed to produce the goods and services which are consumed, say, in a city. The city's 'footprint'[53] is not just its physi-

52 Borgström, G (1967) *The Hungry Planet* Macmillan, New York; Odum, EP (1975) *Ecology: The Link Between the Natural and Social Sciences* second edition, Holt-Saunders, New York

53 Rees, W E and Wackernagel, M (1994) 'Ecological footprints and appropriated carrying capacity: measuring the natural capital requirements of the human economy'. In Jansson, A-M, Hammer, M, Folke, C and Costanza, R (eds.) *Investing in Natural Capital: The Ecological Economics Approach to Sustainability* Island Press, Washington

cal area but includes the life support area provided by nature. The footprint needed to support the residents of the 30 largest cities in the Baltic Sea drainage basin with inputs from fisheries, forestry and agriculture have in one study been estimated as being an area approximately 200 times larger than the cities themselves (Figure 9).[54] The same reasoning is true for the production of individual goods. For example, cage-farmed salmon need a life support area from the environment which is approximately 50,000 times larger than the area of the netpens in which salmon are farmed.[55] Semi-intensive shrimp production in ponds needs an area about 200 times larger than the area of the pond (Figure 10).[56] The shadow areas are often outside the country in question.

Japan and many other nations with a concentrated population and intensive industrial activity would not be able to maintain their level of production and material standard of living through their own natural resources and indigenous ecosystems.[57] They are particularly dependent upon shadow areas which exist far beyond their national borders.

Sweden offers an example of the dependence of foreign shadow areas in its trade with fish and fish products. Sweden is a net importer of fish products in monetary terms and it is therefore asserted that Sweden should decrease its import and concentrate more on building up its domestic fishing industry. The ecological and economic relationships between Swedish import and export of fish products are summarized in Table 1.

Table 3 Relationship between Swedish imports and exports of fish products, 1986[58]

	Import/export
Money (SEK)	2.5
Purchased goods (ton)	2.8
Fish biomass caught (ton)	9.1
Shadow area (km^2)	8.3

The table shows that Sweden imports 2.5 times more fish than it exports, measured in monetary terms. The results also show that Sweden utilizes more of life support systems (8.3 times larger shadow area) per Swedish crown of

54 Folke, C, Larsson, J and Sweitzer, J (1996) 'Renewable resource appropriation by cities'. In Costanza, R and Segura, O (eds.) *Getting Down to Earth: Practical Applications of Ecological Economics* Island Press, Washington, in press
55 Folke, C (1988) 'Energy economy of salmon aquaculture in the Baltic Sea' Environmental Management 12:525–537; Folke, C and Kautsky, N (1989) 'The role of ecosystems for a sustainable development of aquaculture' *Ambio* 18:234–43
56 Larsson, J, Folke, C and Kautsky, N (1994) 'Ecological limitations and appropriation of ecosystem support by shrimp farming in Colombia' *Environmental Management* 18:663–76
57 Odum, EP (1989) *Ecology and Our Endangered Life-Support Systems* Sinauer Associates, Sunderland, Massachussets
58 Hammer, M (1991) 'Marine Ecosystems Support to Fisheries and Fish Trade'. In Folke, C and Kåberger, T (eds.) *Linking the Natural Environment and the Economy: Essays from the Eco-Eco Group* Kluwer Academic Publishers, Dordrecht

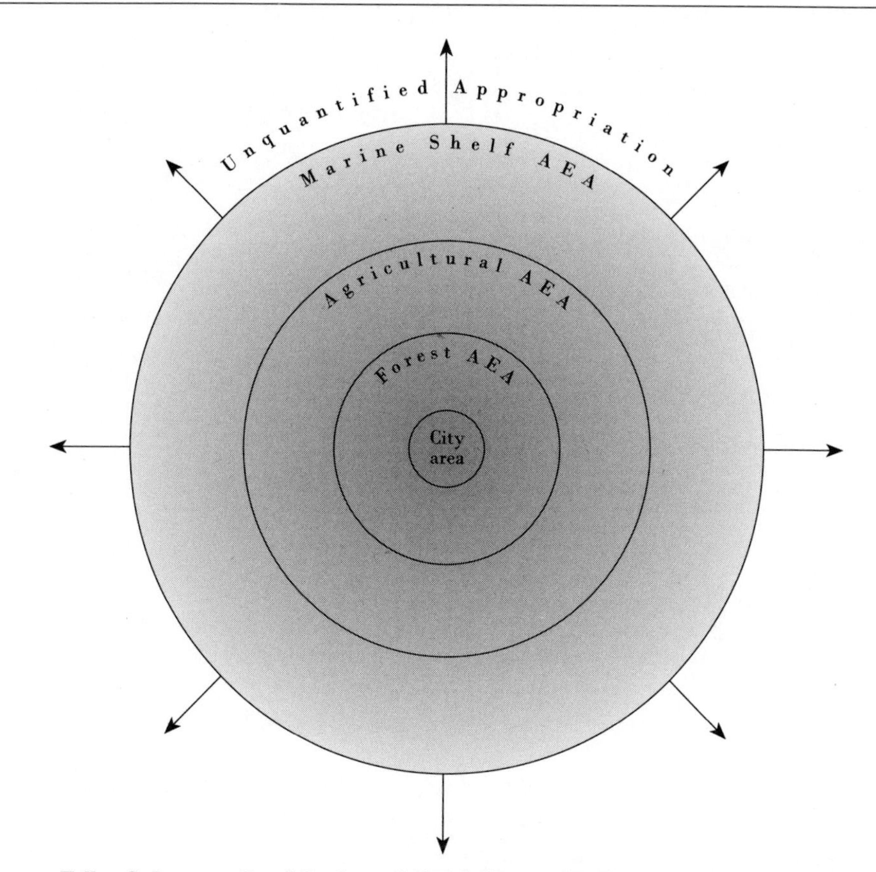

Source: Folke, C, Larsson, J and Sweitzer, J (1996) 'Renewable Resource Appropriation by Cities' In Costanza, R and Segura, O (eds.) *Getting Down to Earth: Practical Applications of Ecological Economics* Island Press, Washington.

Figure 9 The ecological shadow area needed to support the largest cities in the Baltic Sea drainage basin with renewable resources from fisheries, agriculture and forestry.

imported products than per crown of exported products. In other words, Sweden indirectly buys ecological support which is provided by the ecosystem in other countries at a considerably lower price than it sells in equivalent ecological work to other countries. This balance of trade is advantageous for Sweden and illustrates a relationship between developed and developing countries which is not uncommon.

The majority of developing countries are net exporters of food, raw material, minerals and fuel to developed countries; raw materials often dominate their total export figures. For example, more than 98 per cent of the export value of Bolivia, Ethiopia, Ghana and Nigeria consists of raw materials, as compared with 24 per cent for the USA and a bare 2 per cent for Japan.[59]

59 French, HF (1993) 'Reconciling trade and the environment' In *State of the World 1993* Worldwatch Institute/Earthscan, London

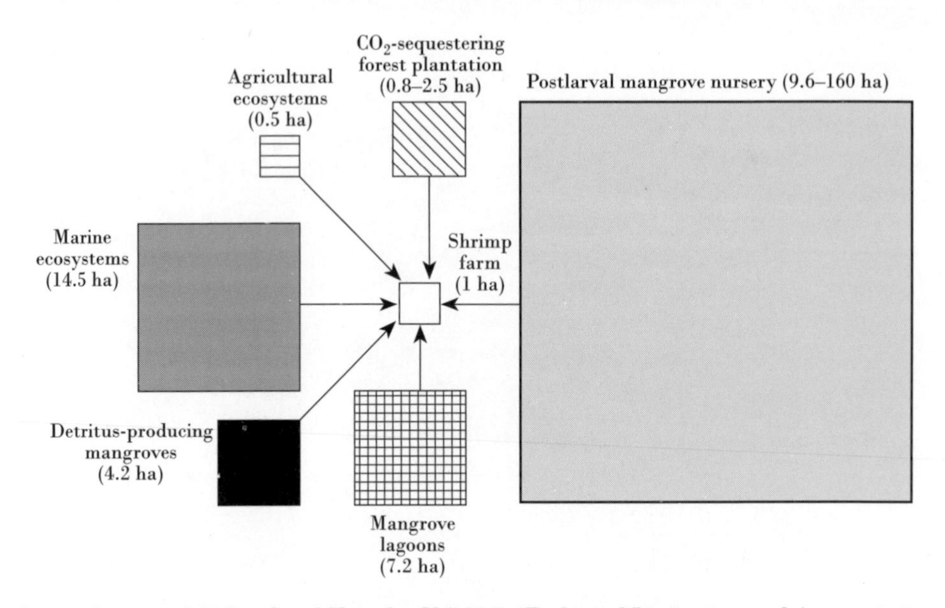

Source: Larsson, J, Folke, C and Kautsky, N (1994). 'Ecological Limitations and Appropriation of Ecosystem Support by Shrimp Farming in Colombia' *Environmental Management* 18:663–676.

Figure 10 The footprint of a semi-intensive shrimp farm in Colombia.

In certain cases, excessive exploitation of natural resources has transformed countries from net exporters to net importers of raw materials. Such is the case for Thailand, Nigeria and the Philippines regarding timber products. The Gold Coast and Ghana will soon be in the same position.[60] Ecological analysis of New Zealand's export of natural resources and of Ecuador's export of intensively cultivated prawns suggests that the countries importing these resources appropriate substantially more of the ecosystem's work than they actually pay for. Export leads to an impoverishment of the export countries' ecological resource base and affects their ability to attain long-term economic stability,[61] in part because the increasing ecological scarcity is not reflected in the export prices.

Ecological services with a high value to society are not reflected by the market prices or the institutions which regulate production, consumption and trade. The monetary flow is mainly related to labour and capital produced by human activity, and to some extent to natural resources and ecosystems with well-defined property rights. But they seldom account for the ecosystem support which is needed in order for trade with goods and services to be at all possible (Figure 11).

60 French, HF (1993) ibid

61 Odum, HT (1984) 'Embodied Energy, Foreign Trade and Welfare of Nations'. In Jansson, A-M (ed.) *Integration of Economy and Ecology: An Outlook for the Eighties* The Wallenberg Symposia, Askölaboratoriet, Stockholm University; Odum, HT and Arding, JE (1991) 'Energy analysis of shrimp mariculture in Ecuador' Working Paper, Coastal Resources Center, University of Rhode Island

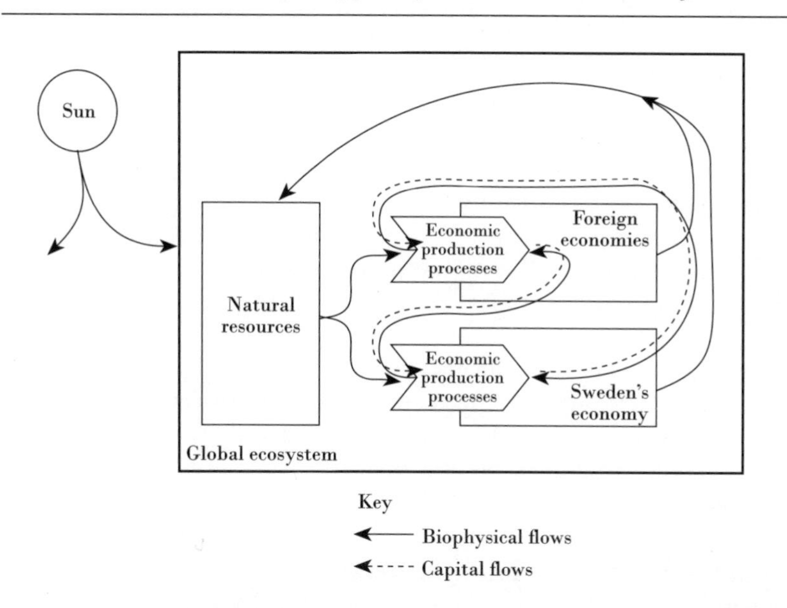

Source: Hammer, M (1991) 'Marine Ecosystems Support to Fisheries and Fish Trade'. In Folke, C and Kåberger, T (eds.) *Linking the Natural Environment and the Economy: Essays from the Eco-Eco Group* Kluwer Academic Publishers, Dordrecht.

Figure 11 Conceptual picture of the economy, trade and the environment. The significance of the environmental resource base for a national economy's welfare falls outside the economic arena.

Specialization leads to vulnerability

Ecosystem resilience can be compared with society's flexibility and ability to adapt to changes.[62] As discussed in Chapter 2, production becomes more specialized with increasing trade sophistication. Specialization results in increased vulnerability and a greater dependence on conditions in foreign nations and their ecosystems, with less opportunity to influence the situation as a result. Widespread specialization can also make it difficult and costly structurally to adjust to novel environmental conditions.

Specialization has often led to throughput-based production systems. Modern monocultures are an example of a type of specialization which is made possible by substantial contributions of auxiliary inputs to sustain production, for example of artificial fertilizer, pesticides, and a high degree of mechanization. Many of these inputs are imported and large shadow areas outside the monoculture region are needed to maintain the system and avoid disturbances. But vulnerability to external disturbances increases. There are many similarities between monoculture and a stressed ecosystem. Both have an imperfect ability to retain energy, nutrients and organic material in addition to being sensitive to attacks by parasites. Recycling occurs on a very restricted scale. Food webs are short, the ecological structures are simple, diversity is reduced and

62 Runnalls, D and Cosbey, A (1992) *Trade and Sustainable Development. A Survey of the Issues and A New Research Agenda* International Institute for Sustainable Development, Winnipeg

the system's efficiency as well as buffer capacity is low.[63]

There is uncertainty concerning ecological thresholds and the capacity of an ecosystem to recover after disturbances. We will always live with uncertainty regarding how to handle complex and interrelated ecological and economic systems. Society should seek to avoid production and consumption patterns which obviously lead to degradation of the ecological resource base, risking irreversible changes. The phrase 'sustainable development' implies that the capacity of ecological systems to support present and future generations should not be reduced.

Vulnerability can be reduced if society works towards completing the ecocycle and actively managing the environmental feedbacks within both production and consumption. Recycling of resources and nutrients, ideally in tune with the ecosystem's processes, creates diversification and resilience while reducing vulnerability.

Trade, transportation and the environment

Without transportation, trade cannot take place and since transportation adds to environmental problems, trade has a direct influence on the environment. It has been estimated that international truck transportation within the EU will increase by 30 to 50 per cent as a result of the development of common markets and deregulation within the transportation sector. Truck transportation in North America is expected to increase seven-fold during the next decade as a result of the North American Free Trade Agreement (NAFTA).[64]

One of the major problems of transportation in the 1990s is that the means of transportation are largely driven by fossil fuels, resulting in air and water pollution. Air and sea transportation of people and goods account for one eighth of the world's oil consumption[65] and are responsible for an already large and increasing proportion of carbon dioxide, nitrogen oxides, sulphur dioxide and hydrocarbon into the atmosphere. Direct oil discharge when transporting oil must be added to this.

Transportation represents a *diffuse* emission source which, in several respects, is more difficult to regulate than emissions from stationary sources, such as specific industries. In practice, it is very difficult to tax aircraft fuel and bunker oil for ships. Such measures must be coordinated by many countries as it would be easy otherwise to fill fuel tanks in countries without a fuel tax. This has resulted in plane and ship fuel being relatively untaxed in most countries, while taxation of fossil fuels for private consumption often amounts to several hundred per cent of the actual market price.

63 Odum, E P (1985) 'Trends to be Expected in Stressed Ecosystems' *Bioscience* 35:419–22; Folke, C and Kautsky, N (1992) 'Aquaculture with its environment: prospects for sustainability' *Ocean and Coastal Management* 17:5–24

64 French, HF (1993) 'Costly Tradeoffs. Reconciling trade and the environment' Worldwatch Institute Paper 113, p.37

65 Madeley, J (1992) *Trade and the Poor. The impact of international trade on developing countries* Intermediate Technology Publications, London

If the total cost of all environmental effects of fossil fuel use were reflected in the price of oil, it would be markedly higher and transportation would be much more expensive. Given such an internalization, not only trade but also production and consumption patterns would be quite different. It would be increasingly profitable to recycle materials and link the recycling to ecosystem processes and functions, in particular at the local level. In addition, the need for transportation dependent on fossil fuel would decrease.[66] The incentive to develop more efficient and environmentally-friendly modes of transportation, for example cleaner boats and trains, would also increase. Today, however, transportation costs are so low that extreme price changes would be needed for such a change to take place (Box 2).

Other environmental effects of transportation are, for example, noise, changes in the landscape due to changed land usage and a 'barrier effect' of roads and railways. In addition, transportation affects the environment via the use of fossil fuel, minerals and other resources when lorries, boats, railways, airports, petrol stations and so on are manufactured or established. The way the transportation system is set up has strategic importance for society's development, with far-reaching effects for the entire community.

Direct and indirect driving forces

Whether trade and the environment are in conflict or cooperate with each other is determined by the institutional structures[67] within and between countries and the various driving forces which cause environmental destruction – economic, social, political, and cultural – that exist within these structures. Direct driving forces can be:

- changes in land usage;
- resource exploitation;
- discharge of harmful elements;
- urbanization;
- industrialization;
- the building of infrastructure.

Indirect driving forces are less obvious and often are often connected to society's institutional structures and the behaviour which these structures promote. Some examples of indirect driving forces are:

- the structures of property rights;

66 Röpke, I (1994) 'Trade, development and sustainability – a critical assessment of the "free trade dogma"' *Ecological Economics* 9:13–22; Günther, F (1993) 'Systemekology och samhällsplanering'. In Berg, PG (ed.) *Biologi och bosättning – naturanpassning i samhällsbyggandet*, Natur och Kultur och Institutet för Framtidsstudier

67 Institutions are constraints devised by humans which shape human interaction. They structure incentives in human exchange, whether political, social, or economic, and shape the way societies evolve through time. North, DC (1990) *Institutions, Institutional Change and Economic Performance* Cambridge University Press

Box 2
ENVIRONMENTAL EFFECTS OF TRANSPORTATION

A study at Uppsala University examined air pollution emission and transport routes for single products. The products were: 1) shirts produced in India, transported via ship from Bombay to Gothenburg via Hamburg and truck to Stockholm; 2) tomatoes flown from the Canary Islands to Stockholm or Rotterdam and then trucked; 3) French bread trucked from Malmö in the south of Sweden to Jokkmokk in the north of Sweden; 4) yoghurt trucked from Le Mans, France, to Stockholm.

Table 4 Emissions of carbon dioxide, sulphur dioxide and nitrogen oxide per transported product

Product	*carbon dioxide*	*sulphur dioxide*	*nitrogen oxide*
1 shirt	222 g	2.9 g	6.4 g
1 kg tomatoes*	1.7 kg	–	5.0 g
1 kg tomatoes**	1.4 g	0.15 g	5.5 g
1 kg bread	144 g	0.23 g	2.7 g
1 kg yogurt	121 g	0.20 g	2.3 g

* Direct air transportation.** Combination of air and truck.

Discharge of sulphur dioxide is especially high for the shirts even though they do not weigh much. This is because the ship fuel is of poor quality and not purified, although it is not difficult to achieve this technically. Boat transport is responsible for 10 per cent of the world's sulphur pollution today. Carbon dioxide emission from air transportation is especially high, because fuel consumption per unit of weight is greater than for other forms of transportation. The 1.5 kilograms of carbon dioxide per kilogram can be compared with the annual global carbon dioxide discharge from fossil fuel, 22 billion tons or just over 4000 kilograms per person per year.

How would charges for atmospheric pollution discharge influence the flow of trade? Environmental charges for individual goods were determined as part of this study and it was assumed that diesel fuel would receive an extra tax of 3.30 Swedish crowns, respectively 10 SEK per litre.

The price of goods is marginally influenced, which implies that trade would be affected very slightly by environmental charges on transportation. This is partly due to transportation costs being very low today, so low that they add an insignificant amount to the total price of the goods. Transportation costs themselves would be drastically affected, which would stimulate transporters to search for environmentally-friendly solutions. However, it is difficult in practice to introduce environmental charges for international transport because if one country has high charges on boat or aircraft fuel, companies will simply fill their tanks where the prices are lower. The need for international cooperation in this particular field is therefore substantial.

Table 5 Environmental charges in relation to the price of products

Product	*Environmental charges (hundredths of a Swedish crown)*	*Charges proportionate to product price*	*Charges proportionate to the cost of transport*	*The cost of transport proportionate to product price*
1 shirt	40 (98)	<1%	30 (80)%	<1%
1kg tomatoes*	88 (260)	4 (10)%	15 (45)%	25%
1kg tomatoes**	77 (230)	3 (9)%	13 (39)%	25%
1 kg bread	18 (55)	0.5 (1.5)%	7 (24)%	7%
1 kg yoghurt	17 (52)	0.6 (1.7)%	no info	no info

The figures within parentheses refer to higher environmental charges.
* Direct air transportation.** Combination of air and truck.

Source: Nycander, G (1992) 'The Environmental Effects of Long Distance Trade', paper written for the Department of Economic History, Uppsala University, Sweden.

- taxes;
- legislation;
- macro-economic policies;
- attitudes of religion;
- worldview of culture.

In Sweden, subsidization of road construction in forests of ecologically sensitive northern mountain areas or *fjell*, has been a decisive indirect driving force for the timber exploitation of these ecosystems. Until a few years ago, land property rights in Madagascar were not well defined. There was therefore little motivation to conserve land, which led to serious soil erosion.

Diamonds and meat are two vital export products for Botswana. Meat production has increased substantially during the most recent decades. Meat production is based on cattle raising on pasture land which has, in turn, caused excessive grazing and a distinct influence on plant and animal life of the savanna. Cattle breeding is subsidized in many different ways by the Botswanan government, and is promoted by the EU which indirectly guarantees to import a certain quantity of meat annually. A total of 45 per cent of export proceeds to the EU (1982) went to commercial cattle breeders, who amount to a bare 0.6 per cent of total cattle breeders. Small local breeders (94 per cent of cattle breeders) received only 33 per cent of the EU proceeds during the same year.[68]

68 Pearce, DW and Warford, JJ (1993) World without End Oxford University Press and the World Bank

Analysis shows that a drop of just a few per cent in world diamond prices has a definite effect on Botswana's diamond exports and thereby on its economy. This makes it more profitable to expand cattle breeding activities, which in turn, leads to even more excessive grazing and desertification.[69]

The clearing of forest in the Amazon has also been accelerated by indirect driving forces. In Brazil, taxes on agricultural income, rules for land allocation, land taxes, regional and sector taxes and the possibility of obtaining loans were indirect driving forces contributing to the loss of the forests.[70] Tax regulations were changed in the late 1980s, but many people who do not own land continue to move to the Amazon area because land reform regulations are not well implemented.

As a result of complex socio-economic relationships, the effects of trade on nature's life support systems are hard to predict. It is therefore difficult to argue that either increased trade barriers or liberalized trade will automatically lead to a sustainable use of life support systems. It is better to concentrate on redirecting the existing institutions that provide the framework for national economies, including trade, so that production and consumption of goods and services are realigned towards sustainability.

If environmental effects are not heeded and trade is opened up, the prospects for export incomes and short-term economic profit increase the temptation to implement unsustainable production. Powerful interests in such situations take advantage of the lack of environmental policy. When this occurs, property rights structures and production systems which have functioned for thousands of years are often completely destroyed.[71]

This type of event occurs today in many coastal areas where mangrove forests are cut down to accommodate intensive shrimp farming, an activity where the product is designated for export. The lost worth of the mangrove forests as a life support system for, among other things, fuel and fish production for a large proportion of the local population, is not included in the market price of the giant shrimp. In such cases, trade raises the need for the internalization of external costs. At the same time, those who profit by such trade activities are highly motivated to fight the internalization process.[72]

Ethics and technology in an internationalized world

Our world view, values, knowledge and institutions influence to a great extent the way in which society relates to nature and the environment. If people believe they rule and are separate from nature then a 'conquering' technology will develop which strives to create a society independent of nature. If people

69 Unemo, L (1995) 'Environmental Impact of Governmental Policies and External Shocks in Botswana: A computable general equilibrium approach'. In Perrings, C, Mäler, K-G, Folke, C, Holling, CS and Jansson, B-O (eds.) *Biodiversity Conservation* Kluwer Academic Publishers, Dordrecht, pp.195–214

70 Binswanger, HP (1989) 'Brazilian Policies that Encourage Deforestation in the Amazon' The World Bank. Environment Department Working Paper no. 16

71 Ekins, P, Folke, C and Costanza, R (1994) 'Trade, environment and development: the issues in perspective' *Ecological Economics* 9:1–12

72 Röpke, I (1994) 'Trade, development and sustainability – a critical assessment of the "free trade dogma" *Ecological Economics* 9:13–22

regard themselves as part of nature and recognize their dependence on its support then a more collaborative type of technology, known as ecotechnology, will develop.

Technologies are not simply tools which can be put to good or bad use; they are a reflection of the cultural values, world view, knowledge and institutions of the society where they are formed. This is known as cultural capital.[73] A simplified illustration of a systems view of the relationship between nature, technical advancement and culture can be seen in Figure 12. If technology masks society's dependence on life support systems and continues to procrastinate on environmental problems, then people will be lulled into believing that

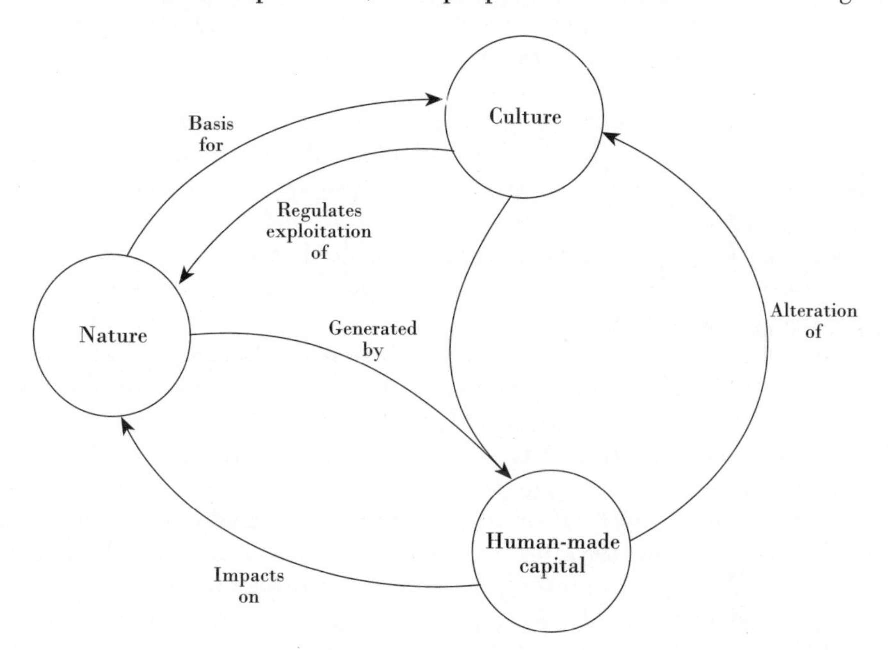

Source: Berkes, F and Folke, C (1994) 'Investing in Cultural Capital for a Sustainable Use of Natural Capital. In Jansson, A-M, Hammer, M, Folke, C, Costanza, R (eds.) *Investing in Natural Capital: The Ecological Economics Approach to Sustainability* Island Press, Washington.

Figure 12 First order relations among nature, capital, and culture. Life support systems are a prerequisite, a necessary but insufficient condition for culture and welfare. Tools and techniques are created by an interplay between nature and culture. Technical developments impact on nature's life support systems, for good and for bad, but also effect the cultural view of humanity's relation with nature in a sustainable or unsustainable direction.

73 Berkes, F and Folke, C (1992) 'A Systems Perspective on the Interrelations between Natural, Human-made and Cultural Capital' *Ecological Economics* 5:1–8; Berkes, F and Folke, C (1994) 'Investing in Cultural Capital for a Sustainable Use of Natural Capital' In Jansson, A-M, Hammer, M, Folke, C and Costanza, R (eds.) *Investing in Natural Capital: The Ecological Economics Approach to Sustainability* Island Press, Washington pp.128–49; Hjort af Ornäs, A and Svedin, U (1992) 'Cultural Variation in Concepts of Nature' GeoJournal 26:167–72

we are 'above' nature and can manage without it. This will lead to institutions basing their policy on a 'conquering mentality', which in turn influences teaching and research, information, problem solving, decision making, aid to developing countries, and so on.

If a country continues to damage its ecosystems, and thereby the goods and services it produces, it implies that people rely on a technology which seems to make it possible to substitute the loss of domestically produced goods and services for similar goods and services imported from other countries. But the scale of human activity continues to increase and environmental damage does not only occur in indigenous ecosystems. Humanity is in a new era of co-evolution of jointly determined ecological and economic systems (see page 17). From the perspective of sustainability, such an exchange is therefore merely an illusion.[74]

People are becoming more and more dependent on and are wielding greater influence – even if it is unconscious – over situations which occur in other countries. This trend is strengthened by increasing trade and environmental effects which are spreading over larger and larger geographical areas. The scope of an individual's influence requires taking larger responsibility, especially in our role as consumer, and there is a need for new norms and ethics in people's actions.

The internationalization of the world economy can, however, result in people having greater difficulty in taking responsibility for their behaviour. An individual's actions can be regarded as a drop in the ocean. One of the challenges of creating sustainable development lies in turning human interest away from the individual in isolation, to a recognition that it is in the self-interest of individuals to contribute to the maintenance of functional life-supporting ecosystems. An increased recognition of our dependence on resources and services produced by ecosystems in other countries would add to the motivation to contribute to the conservation and sustainable use of those ecosystems. For this to occur, we require more information about fundamental ecological and socio-economic interdependencies and an inner desire to connect remote occurrences with our own actions.

At the same time, experience suggests that ethical rules for sustainable behaviour develop when people are directly confronted with the consequences of their actions.[75] Society can partially speed up this process by introducing different forms of policy. One example would be prohibiting the export of hazardous waste to other areas and countries. By removing the possibility of shipping away all the unpleasant consequences of consumption, people would be forced to tackle the problems they have generated themselves or prevent them from occurring in the first place.

History shows that many societies which are dependent on local ecosystems have developed sophisticated ecological knowledge with social and ethical

74 Folke, C, Hammer, M and Jansson, A-M (1991) 'Life-support value of ecosystems: a case study of the Baltic Sea Region' *Ecological Economics* 3:123–37

75 Röpke, I (1994) 'Trade, development and sustainability – a critical assessment of the "free trade dogma"'. *Ecological Economics* 9:13–22

rules guiding sustainable resource use. This knowledge and relationship to life support systems have developed as a necessity. These societies have learned how to handle feedbacks from the ecosystem in order to survive. The global trend towards sustainable development can be seen as an international effort to handle feedbacks from the global ecosystem.[76]

Many local societies have knowledge of how to use resources and ecosystems in a sustainable fashion. The shift from local resource dependence to globalization can result in the loss of such traditional ecological knowledge and of the institutions and property rights which preserve knowledge of a sustainable use of the ecosystems.[77]

> *As traditional peoples are integrated into the global economy, they lose their attachment to their own restricted resource catchments. This could lead to a loss of motivation to observe social restraints towards the sustainable use of a diversity of local resources, along with the pertinent indigenous knowledge that goes with it.*
>
> Berkes, F, Folke, C and Gadgil, M (1995) 'Traditional Ecological Knowledge, Biodiversity, Resilience and Sustainability', in Perrings, C et al. *Biodiversity Conservation: Policy Issues and Options* Kluwer Academic Publishers, Dordrecht.

76 Berkes, F, Folke, C and Gadgil, M (1995) 'Traditional Ecological Knowledge, Biodiversity, Resilience and Sustainability' In Perrings, C, Mäler, K-G., Folke, C, Holling, CS and Jansson, B-O (eds.) *Biodiversity Conservation* Kluwer Academic Publishers, Dordrecht, pp. 281–299

77 Gadgil, M (1987) 'Diversity: cultural and biological' *Trends in Ecology and Evolution* 2:369–73

Chapter 2

Economic Perspectives on Trade and the Environment

Interest in the link between trade and the environment has grown quickly during recent years and the literature on the subject is expanding rapidly. The character of environmental values was discussed in Chapter 1; this chapter will present a short summary of the economic perspective of trade. We then focus on the connection between trade and the environment. The environment's international aspects are examined and possible ways of handling problems through harmonization and international cooperation are considered. The chapter further discusses the usefulness of trade barriers as an instrument for handling environmental issues, the localization of activities which damage the environment, and the political factors which might contribute to difficulties in this field.

THEORIES OF TRADE

Classical theories explain international trade of goods as a result of countries having varying conditions, or comparative advantages, for different types of production. According to Heckscher-Ohlin, a country's comparative advantages are determined by the availability of production factors relative to other countries and the input that is required within different industries.[78] A country rich in labour power but with little capital has a comparative advantage in labour-intensive production. That country will most likely export goods such as clothing, and import capital-intensive goods such as machinery. An alternative explanation of comparative advantages, originally developed by Ricardo-Viner, stems from differences between countries' technology. Five types of resources are often identified: natural resources, real capital, labour,

78 Heckscher, E and Ohlin, B were active at the Stockholm School of Economics during the respective periods 1909–44 and 1929–65

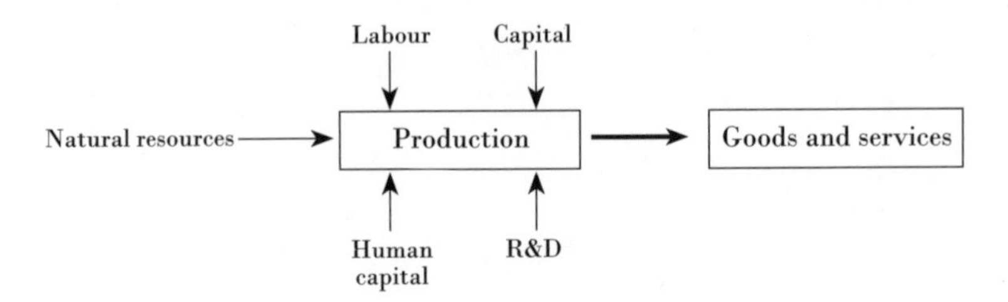

Figure 13 The five factors of production.

capital in the form of labour–knowledge (human capital) and the ability to produce new knowledge through research and development (R&D) (Figure 13). We stress the environment – nature's life support systems – should be regarded as an additional resource.

By facilitating specialization of production across national boundaries, it has been shown that trade leads to the highest possible welfare in the participating countries. The classical models were, however, based on a number of assumptions which are seldom or never fulfilled in reality. For example, it was assumed that all costs and revenues are incorporated by the market, that perfect competition exists among all players – neither single countries nor companies are able to influence prices – and that there are no economies of scale in production.

With time, it has become clear that trade theory in its traditional form is not sufficient. It has since been complemented and revised by newer theories, although the specialization of production in accordance with comparative advantages remains an important tenet of trade theory. The bulk of international trade occurs today within the same industries, which results in countries buying and selling similar products. Trading partners are quite similar with regard to access to resources and technological capabilities. Therefore, the intensity of exchange cannot simply be explained by differences between countries.[79]

In addition, a significant proportion of international trade in goods and services occurs within individual companies – often referred to as multinational companies. These control foreign subsidiary companies through direct investments, meaning that they have sufficiently large shares to establish a substantial, long-term influence. Direct investments would not be possible if perfect competition prevailed in all markets, but presuppose access or abilities specific to the individual organization.[80] Multinational companies transfer competence within, for example, production and distribution, and the over-

79 This was noted by Linder, SB (1961) *An Essay on Trade and Transformation* New York, Wiley. He explained these types of trade patterns on the principle that similar countries find it easier to adapt to each other's requirements.
80 Hymer, SH (1960) The International Operations of National Firms: A Study of Direct Foreign Investment, Ph.D. dissertation, MIT. See also Dunning, JH (1977) 'Trade, Location of Economic Activity and the MNE: A Search for an Eclectic Approach' In Ohlin, B, Hesselborn, P-O and Wijkman, PM (eds.) *The International Allocation of Economic Activity: Proceedings of a Nobel Symposium Held at Stockholm* Macmillan, London, pp. 395–418

Box 3
THE FOUNDATIONS OF TRADE THEORY

Classical trade theory still constitutes the point of departure for most economic analysis of trade, and above all guides trade policies as well as other areas of decision-making. Trade theory explains why trade exists and leads to increased material welfare for all trading partners. This is the key argument for liberalizing trade. Proponents of free trade argue that since increased trade leads to increased prosperity, it should be restricted as little as possible.

Assume that a German worker can produce two spades or five pairs of shoes per hour and a Russian worker can produce one spade or ten pairs of shoes. Both profit in this situation by specializing in the activity at which they are best and then trading with the other. The German specializes in spades and the Russian in shoes, the type of production in which they each have an absolute advantage. This example illustrates a fundamental principle within economic theory; specialization results in higher total production.

Furthermore, however, trade theory shows that countries which do not have absolute advantages in some type of production will also profit from trade. If a German worker can make both shoes and spades cheaper than the Russian can, why should he buy shoes from the Russian? The answer is tied to the concept of 'comparative advantages', coined by the English economist Ricardo in the early 1800s. Perhaps the German produces both shoes and spades more efficiently than the Russian, but presume that the difference in efficiency is greater for the spades. Germany then has comparative advantages in spades and Russia has comparative advantages in shoe production. Production of a good in which a country has a comparative advantage leads to increased specialization and greater total production.

This simple model with just two goods and two countries can be generalized and used for many goods and countries without a change in the final result: each country is assumed to have comparative advantages in some type of production, which will result in all countries profiting from trade.

Ricardo assumed that comparative advantages were caused by countries using different production techniques. According to the Heckscher–Ohlin model, which is still the fundamental trade theory, it is rather the interplay between different factors of production which determines comparative advantages. Countries with an abundance of labour are presumed to specialize in labour-intensive production, while countries which possess more land in relation to the work force have comparative advantages in land-intensive production. Naturally, the international division of labour is not complete. Various aspects, for example transportation costs and cultural differences, are disregarded.

These theories of trade are built upon certain assumptions which are seldom or never fulfilled in reality, such as perfect competition, that goods and production resources are correctly priced and that factors of production cannot move between countries.

whelming share of all international technology flows today occurs within their organizations.

According to recent theories, trade patterns are primarily a result of imperfect competition, asymmetric information, economies of scale, and variation in the design of products.[81] When attention is given to such factors, it becomes clear that trade can be harmful under certain conditions. On the other hand, trade has potential advantages in addition to those considered by classical theorists, namely, trade makes it possible for companies to gain more from economies of scale, increased competition makes it difficult to charge high prices, and customers can be offered a more varied selection of goods and services.

Trade policy

Trade and its effects on welfare are determined not only by the market but by political institutional factors. This can occur overtly, for example when the framework and regulations for trade are decided, or as a result of unforeseen side effects. Policy measures can compensate for market failures but may also hamper the market. Although those who promote political interference have received support from 'strategic trade theory', the mainstream economic view of trade continues to be stamped by the insight that clear and open groundrules are in the best interest of all parties.[82]

Trade policy measures can usually be divided into customs' duties and non-tariff trade barriers, of which the latter cover a highly varied collection of measures. Besides quotas and other import regulations, non-tariff barriers consist of product norms, subsidies and other measures which although not taking the form of border barriers, still restrict foreign sales. While customs' duties have been steadily reduced since World War 2 due to repeated negotiations under GATT, the non-tariff trade barriers have changed character. In particular, 'voluntary export restraints' and anti-dumping measures have emerged as instruments which can effectively restrict exports from specific countries and companies.

In the next chapter, we will discuss in greater depth trade policy systems and their connection to environmental questions. It should be noted, however, that multilateral trade negotiations have only been able to advance slowly while the bulk of international trade has been extensively liberalized on a regional level, most significantly within the EU and NAFTA. The United States has systematically implemented bilateral trade agreements and has

81 Product differentiation can be motivated by consumers' valuation of product variation, or by consumers having different preferences. See: Krugman, PR (1979) 'Increasing Returns, Monopolistic Competition and International Trade' *Journal of International Economics*, 9, 469–79; Lancaster, KJ (1979) *Variety, Equity and Efficiency* Columbia University Press; Helpman, E and Krugman, PR (1985) *Market Structure and Foreign Trade* MIT Press, Cambridge, Mass. For an overview of more recent trade literature, see Grossman, GM (1992) (ed.) *Imperfect Competition and International Trade* MIT Press, Cambridge, Mass. For an overview of costs for restrictions of trade, see Feenstra, RC (1992) 'How Costly is Protectionism' *Journal of Economic Perspectives*, 6, pp. 159–78

82 The dangers of strategic trade policy have been emphasized by, among others, 40 US economists in 'Trade Policy' *World Economy* 2, pp. 263–6

forced various countries to agree to 'voluntary export restraints'. The EU also follows this line, with the result that it has become more important for small countries not to stand alone in international negotiations. There is a widespread anxiety regarding the emergence of 'trade blocs', which while liberalized from within also set up barriers against each other. At the same time, many observers feel that regional reduction of trade barriers paves the way for more open trade worldwide.

Transformed trade patterns

As shown in Figure 14, industrial products have regained and even overtaken the dominant position in trade which they enjoyed before the oil crises. At the same time, industry has been reoriented towards more complex high technology products, and services have become a central component of nearly all production processes. That the value of trade in services has risen far above trade in goods is, however, misleading since financial flows move back and forth within seconds without any visible transfer of resources taking place.

The ability to handle knowledge and modern technology therefore provides the basis for an increasing proportion of the world's material production and trade, while natural resources and inexpensive labour power account for a diminishing proportion when analysed in economic terms.

Parallel to this course of events, which has involved an intensive growth of trade and investment within the triad of North America, Western Europe and Japan, the developing world maintained a marginal role in the global economy

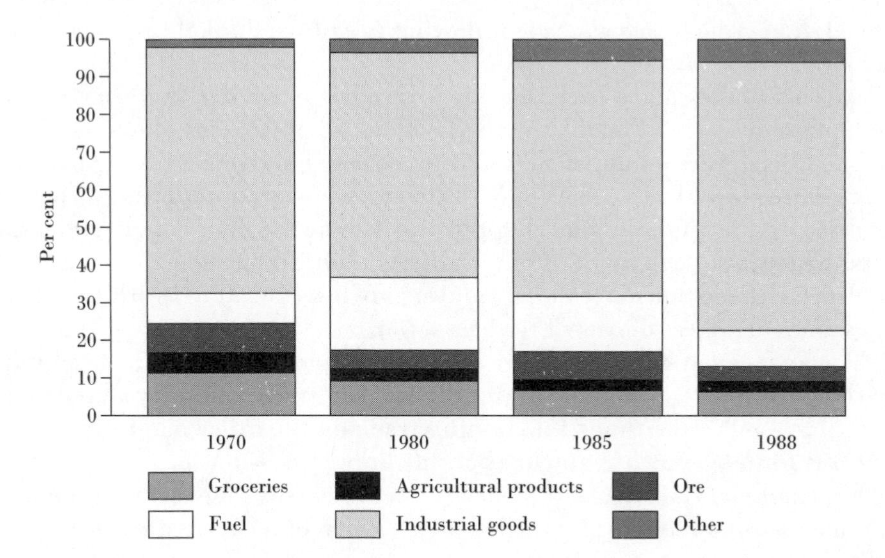

Source: UNCTAD (1987 and 1990) *Handbook of International Trade and Development Statistics* New York.

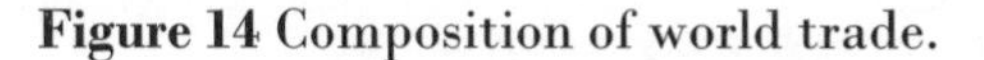

Figure 14 Composition of world trade.

during the 1980s, despite the fact that the majority of the world's population and natural resources are located there. The situation is changing in the 1990s. The growing importance of the newly industrialized countries in East Asia, where success has sharply intensified competition within industry, has been obvious for years. Today, other parts of East Asia, including China and Vietnam, Latin America and even South Africa are attracting increased attention. The traditionally powerful economies of North America and Western Europe have lost their grip on the fastest expanding segments of world trade, and are seeking new ways to respond. The former planned economies of Eastern Europe have dropped back even more drastically, but there is hope that some of them can soon enter the group of successful, newly industrialized countries.

Other forms of trade

Trade not only consists of a flow of goods and services that are exchanged between buyers and sellers in the international market place, but a large part of trade occurs within countries and is regulated by 'national trade barriers'. For example, various functional regulations, formal educational requirements, and so on act in reality as non-tariff trade barriers which influence international trade. While barriers at the border may give rise to smuggling, domestic trade barriers tend to result in tax evasion and growth of a black market. Furthermore, the line between national and international trade is blurred, as trade between regions in a country may be subject to the same limitations as trade between different countries. With the creation of customs' unions, single markets and currency unions, the decision-making responsibility for forming the regulations which govern trade is moving from the national to the international level.

Trade also takes place over time, as is implicit in saving, investment and consumption decisions. Finally, there exist flows which are entirely or partially non-commercial, which implies that well-functioning markets do not exist. The result is that trade fails to come about, although transfers of resources do occur. As is discussed in the previous chapter, this is true for many environmental effects, which take the form of externalities and fall outside the market.[83] Under such circumstances, economic players are not confronted with the actual costs and outcomes connected to their actions.

The management of different values can be divided on scales which contain everything from perfect markets to totally blocked trade on the one hand, and transfer of resources without functioning trade on the other hand. Changing conditions for one type of value may result in repercussions on the others as well. For example, the effects of abolishing trade barriers for certain products are decided to a high degree by the extent to which other related resources are handled by functioning markets.

When costs and revenues are not all incorporated into the market, many

83 Pigou, AC (1932) *The Economics of Welfare* Macmillan, London

Box 4
ENVIRONMENTAL SUBSIDIES AND ELECTRICITY PRICES IN THE SOUTH

If environmental costs are not reflected in prices, it indicates that production is subsidized. Determining the extent to which a specific product has been subsidized is extremely difficult, however. It is impossible for other countries to say exactly how strict the environmental standards of an exporting country should be.

On the other hand, it can be stated that many products and processes are subsidized even when environmental costs and benefits are disregarded. Electricity prices – which are central to environmental policies – are subsidized in most countries, and particularly in the South. The price of electricity in the developing world is, on average, half what it is in industrialized countries. In India and China it is even lower than that. From 1979 to 1988, real prices rose by 1.4 per cent in OECD countries while they fell by 3.5 per cent in the South.

A basic principle of economic price theory is that price should be based on marginal costs, which means it should be equal to the cost of producing the last unit of an item. If prices are lower, over-consumption is stimulated, while other sectors are negatively affected. On average electricity prices in the South represent 60 per cent of the marginal costs of production, which means that too much capital is spent on energy-demanding projects. In addition to financial subsidies, in the South we have the implicit subsidy of environmental effects.

When electricity prices are low, saving energy becomes less attractive. On average, 20 to 40 per cent more fuel is used to produce a kilowatt hour in the South than in the North. Between 15 and 20 per cent of the power that is produced disappears because of theft and losses during transmission. This all contributes to a situation where 40 per cent more energy is required for production in the South than in the North. Governments and development organizations have emphasized increasing production rather than efficiency. Of the loans made to the energy sector from the World Bank, it has been estimated that less than 1 per cent have been used to raise efficiency. Loans for production of electric power account for 20 to 40 per cent of the developing world's foreign debt.

Many investments in electricity generation have been made in obsolete technology, especially in large-scale hydropower projects and coal-fired plants, which account for 50 per cent of developing countries' electricity production. China installs a new coal-powered plant with a capacity of 1000 megawatts every month. Higher electricity prices would encourage implementation of superior alternatives, such as more effective technology, natural gas and small-scale hydropower projects. For political reasons, however, it is difficult to raise prices. It might be easier to invest in efficiency directly so that production costs are matched by prices. The fact that energy production is so inefficient today means that there are great opportunities for future savings.

Source: 'Power for the poor', *The Economist* 3 August 1991; World Development Report 1991, the World Bank; *State of the World 1993* Worldwatch Institute.

predictions of trade theory are no longer accurate. In order to correct for the occurrence of external effects, either functioning markets must be created, or charges and regulations must be introduced to compensate for the lack of functioning markets. The goal is that costs and revenues become the responsibility of those who cause them; the externalities are internalized.

Evaluation of the environment

Even though it may be difficult in many cases to evaluate the importance of life support systems to society, a cost–benefit analysis can be used to estimate society's priorities and thereby get away from the attitude that environment degradation does not involve an actual cost. Without getting overly involved in the principles of this type of evaluation,[84] it can be stated that the best economic solution is reached when consideration for the environment is demanded to the degree that expected profits are as large as the costs *at the margin*. The 'first best measure', whether it is a tax or a market for emission rights, forces those who are responsible for damage to bear the actual cost and thereby stimulates technological development for improved environmental management. In certain situations it is preferable to regulate, for example in order to simplify compliance control.[85]

It should be emphasized that investments in the environment and commercial activities do not need to conflict with each other – to represent *substitutes*. In many cases, they actually enhance each other – they constitute *complements*. The same efforts may serve as substitutes in one situation and as complements in another.

The principle of substitution largely dominates in the short-term perspective. For example, an investment in purification equipment can lead to a company having to abandon other expansion plans. Complementarity tends to appear stronger in the long-run, since environmental concerns may give rise to technical progress, and since the environment functions as an infrastructure for continued activity while many environmental effects evolve gradually. A similar logic applies in terms of spatial or geographic dimensions. Environmental damage often has long distance effects while the cost of pollution reduction is paid locally; complementarity is generally stronger the larger the area involved. For example, preserving rain forests benefits the entire world, while the costs tend to be borne by individual countries and are more concentrated on the people, often very poor, living in the immediate vicinity of the rain forest.[86] This implies that investment in the environment and a better material standard of

84 Little, IMD and Mirrlees, JA (1974) *Project Appraisal and Planning for Developing Countries* Heinemann, London; Helmers, FLCH (1979) *Project Planning and Income Distribution* Martinus Nijhoff Publishing, Boston, the Hague and London; Bojö, J, Möler, K-G and Unemo, L, (1990) *Environment and Development: An Economic Approach* Kluwer Academic Publishers, Dordrecht

85 Cropper, ML and Oates, WE (1992) 'Environmental Economics: A Survey' *Journal of Economic Literature*, 30, pp. 675–731

86 See Wells, M (1992) 'Biodiversity Conservation Affluence and Poverty: Mismatched Costs and Benefits and Efforts to Remedy them' *Ambio*, 21, pp. 237–43

living are more likely to be regarded as complementary the greater the consideration is for future values, and the more that different regions and countries work cooperatively towards a common goal.

Individual players are often tempted to ignore complementarity, to the extent that they can profit by their actions without bearing the entire cost. On the other hand, there is a growing tendency for consumers to appreciate environmentally-friendly products and production processes and companies can profit from 'goodwill' by taking advantage of this. Governments also have a particular responsibility to promote consideration of the environment, since individual players cannot be expected to consider effects as long as the profits are partially reaped by others.

TRADE AND THE ENVIRONMENT

Keeping in mind the changing role of the nation state, we may still generalize by saying that the world consists of sovereign states which have their own territories but share a common resource base. In practice, it can be difficult for a single country to introduce taxes or other measures, even if the 'actual' environmental costs are known and there are opportunities to control the situation. The reasons for this are rooted in the international aspects of the environmental problem.

> *The 'spider's web' phenomenon (global integration) has meant increasingly that everyone tends now to be in everyone else's backyard, making import competition in one's own market, and export competition in the other's market and in third markets, ever more fierce in an atmosphere reminiscent of the struggle for the sun in a dense tropical forest.*
>
> Bhagwati, J (1991) *The World Trading System at Risk*, Princeton University Press, p.16.

A socially wasteful degradation of natural capital is caused by a number of driving forces, such as production and consumption patterns governed by, for example, insufficient information and a lack of well-defined ownership rights. The absence of a functioning market for environmental effects and political measures decided by governments reaping short-term profits from exploitation of the environment, create an incomplete framework for more or less all economic interactions. The connection to trade is obvious, but has only been highlighted recently. In principle, there is no conflict between liberalized trade and consideration to the environment since both have the same goal; to use and distribute the resources available to society in the most efficient manner. Trade in itself poses no problem, as long as effects are internalized, meaning that those who cause problems bear the costs and benefits associated with them.

Trade in this case makes it possible for society to achieve the highest possible gain. Free trade in commercial goods alone, however, where environmental effects are not taken into account, comprises a completely different situation.

The effect of trade on the environment

The effect of trade on the environment is interpreted differently depending on the point of departure. It can be said that trade, by preparing the way for specialization, facilitates higher income-levels and faster technological development, which create more resources for environmental investment and a greater appreciation of environmental values. On the other hand, it can be argued that expanding transport and production increases pressure on the environment. Most modern agriculture, for example, with its utilization of heavy machinery, chemical insecticide and artificial fertilizer, eliminates competing biological resources. This has already occurred in developed countries and is now taking place in developing countries. Losses occur partly because diversity is in itself a resource which has value – it is more desirable to have many as opposed to fewer alternatives – and partly because diverse resources, each one with a varying yield, increase society's ability to adapt to changes.[87]

Without touching upon all the connections between trade and the environment, it can be concluded that it is not very meaningful to debate which of these arguments weighs the heaviest. In a situation where environmental effects are not taken into account, it cannot be determined *a priori* whether trade will promote or damage the environment. But it can be stated that trade strengthens economic development whether it is sustainable or not.

International environmental problems

Simply put, there are two basic causes of international environmental problems. First, activities in one country may affect the environment in other countries. Second, activities which cause environmental damage can move abroad. Assuming that each nation state is sovereign and that governments are concerned only with the welfare of their respective citizens, individual countries may for both these reasons be less motivated and have fewer opportunities to implement international environmental demands within their territories. For example, the total environmental damage can increase if a single country implements stricter measures, since others may thereby become less motivated to take any action.[88]

Let us consider the simple case in which one country may cause environmental damage in another country as a result of the environment's transnational dimensions. Three possible effects are illustrated in Figure 15:

87 Swanson, T (1991) 'Biological Diversity as Insurance' Center for Social and Economic Research in the Global Environment, Research Paper 92–04, SCERGE, London
88 Hoel, M (1989) Global Environmental Problems: The Effects of Unilateral Actions Taken by One Country, mimeo, University of Oslo

Box 5
TRADING IN WASTE MATERIAL

During the latter part of the 1980s, trade in environmentally hazardous waste started to attract attention. Waste is exported from one country to another for final disposal (dumping) or recycling. For the most part, this trade occurs between industrialized countries, but the trade that has received the greatest attention is exports to Eastern Europe and Africa. There are no reliable statistics on just how comprehensive this activity is, since it is often quite 'shady' and occasionally even illegal. According to Greenpeace, at least 10 million tons of waste of all kinds have been exported during the last few years; more than half to Eastern Europe or developing countries. According to Logan, the same amount was imported by Africa, south of the Sahara, during 1987.

One reason for the increase in this type of trade is that the cost of waste disposal rises as the amount of waste increases and environmental protection laws tighten up. It becomes profitable to export the material where it can be disposed of inexpensively. In principle, there is no difference between trade in waste and trade in goods which result in environmental effects when they are consumed. On the other hand, waste management may contain elements of all three kinds of international environmental effects, depending upon whether waste is exported or dumped in the original country without satisfactory processing and how local the resulting damage is.

In early 1992, an internal memorandum written by Lawrence Summers at the World Bank was leaked to the press. In the paper, which spread rapidly and received enormous attention, Summers suggested that the World Bank encourage the export of waste to developing countries and that environmentally damaging businesses and activities be moved to the same countries. He believed that the costs for environmental damage would be lower in the South than in the North for three reasons:

- Costs in the form of production and income losses in cases of illness or premature death are lower in the South, since average life-span and incomes are lower.
- Those countries which are still not polluted have a larger capacity to assimilate toxic waste than countries in the North, where pressures on the environment are already sizeable and even marginal additions of pollution can be extremely costly.
- The demand for a clean environment for aesthetic and health reasons has a 'lower priority' in poor countries, and therefore costs are not evaluated as highly when the environment is damaged.

With the first point, Summers states that people's lives should be evaluated in terms of their incomes. This is in opposition to the UN declaration of human rights, which states that all individuals have a right to life, regardless of income. When analysing the value of building highways, individuals' time is sometimes evaluated differently, according to differences in income. Saving time travelling on roads which approach airports is, for example, rated particularly high. To rate lives in the same way is another matter.

The two other points are a common application of economic theory. According to this perspective, prices and costs are decided by people's values. People's preferences (priorities) are influenced by, among other things, income; it is often assumed that when incomes rise so do priorities for the environment. In this case, it means that a poor population is assumed to evaluate the costs and risks of waste disposal lower than a wealthy population would. The 'price' of the environment is therefore lower.

Instead of, as Summers has done, using the value of the environment or people's lives as a departure point, we can examine how *risk calculation* is determined in different countries. It can then be established that the risk of death is rated lower in poor countries, so waste disposal is naturally cheaper there. If the cost of waste disposal varies, all countries profit from localization in the area where it is the lowest. The theoretical result will then be that trade in waste should be encouraged and accepted. This does not mean, however, that trade in waste as it is now should be encouraged. Several factors which are not heeded in theory, such as institutional relationships and geographic conditions, support the argument that a great deal of today's waste trade is not desirable, as follow:

- Biological damage resulting from toxic waste can be greater in Africa than in Europe or in the USA. The natural *geographical conditions* are such that the risk of poisonous materials spreading over large areas via ground and surface water supplies is much higher. The higher precipitation in countries south of the Sahara easily leaks poisonous elements from deposits, and in the Sahara area poisons infiltrate quickly through the sandy soil to the ground water.
- In theory we assume that both partners in a transaction have complete information in order to arrive at rational decisions, but such is not the case with the waste trade where there is a chronic *lack of information:*
  – The effects on the environment are often not known, especially in tropical climates. The general uncertainty regarding future damage makes it difficult to weigh the costs of trade against incomes, while as a rule only the latter appear directly and are calculated financially.
  – The different parties do not have equal access to information. Importers often have less experience in waste disposal than exporters and do not receive information regarding the specific contents of the waste they will be dealing with.
- The volume of toxics is vast and increasing. Trade causes hazardous transportation, since increased transportation of toxics increases the risk of accidents or even malversation, particularly of radioactive substances.

- Perhaps the strongest argument against waste trade is based on the balance of power within importing countries and how they function in reality.
  – Incomes from imports seldom go to those exposed to risks, and since import countries are generally not democracies the population in question virtually never has a say in import decisions.
  – Import countries in the South often do not have the political and institutional stability required to manage waste disposal in a long-term secure manner, and the authorities often do not have the capacity to control imports or discover and penalize illegal imports.
- Low prices hamper development of technology as lack of information leads to dramatic variations in price. The costs of waste disposal in the North are US$160 to US$3000 per ton, depending on the level of toxicity. The average payment for dumping waste in Africa is US$2.50 per ton (Logan). With such low costs, the countries which produce waste are less motivated to develop technology to manage their own waste disposal.

The Basel Convention on exports of waste

A convention on transboundary movements of hazardous waste was approved in 1989 and came into force in May 1992 after it had been ratified by 20 countries. Today, 30 countries are party to the convention, which requires exporters to obtain prior consent from the receiving nation before approving cargoes. Developing countries want to stop all waste export, but the only movement of waste forbidden by the treaty is to the Antarctic.

A cooperative conference for the convention's various partners was held in December 1992. UNEP suggested a total ban on exports from the North to the South, which the USA, Canada and England opposed. The South considered their own import ban difficult to control, and therefore preferred a ban on exports by the North. The compromise which was adopted at the time meant that export for recycling or re-use was allowed, although it is not absolutely clear what the limits to recycling really are. (The Walloon ruling from the EC Court of Justice, see Chapter 3, demonstrates that definitions of waste and raw materials are not self-evident.) The substantial profits from trade meant that certain groups had a strong interest in preserving the status quo. Still, in March 1994 a total ban on exports – for disposal as well as recycling – from the North to the South was agreed upon.

Sources Daly, H E and Goodland, R (1994) 'An Ecological–Economic Assessment of Deregulation of International Commerce under GATT' *Ecological Economics* 9:73–92: French, H F (1993) 'Costly Tradeoffs. Reconciling trade and the environment' Worldwatch Institute Paper 113, p.25; Logan, B I (1991) 'An Assessment of the Environmental and Economic Implications of Toxic Waste Disposal in Sub-Saharan Africa' *Journal of World Trade* 25:61–76; 'Let them eat pollution' *The Economist* 8 February 1992; 'Pollution and the poor' *The Economist* 15 February 1992.

Box 6
MORE THAN TOBACCO GOES UP IN SMOKE

Tobacco is one example of a traded good that has negative environmental effects both during production and consumption. The effects of production are not discussed very much, but in some areas they are serious. When tobacco is harvested, it is often dried before being further refined. Some varieties can be cured in the sun, but those that are most often grown for export require the use of fuel, either fossil fuel or firewood. Wood is used to cure half of the world's tobacco, mainly grown in developing countries. The process is inefficient – only 20 per cent of the heat is used – so large amounts of firewood are needed. It has been estimated that one tree is needed for 300 cigarettes, or that an acre of tobacco requires an acre of forest. Large areas are involved in the total picture; the import of tobacco to England, for example, consumes almost 200,000 hectares (490,000 acres) of forest per year.

In total, 35 million small farmers all over the world are dependent on tobacco cultivation. Tobacco exports are a significant source of income in Tanzania, Brazil, Sierra Leone and Mali. Tobacco export accounts for 25 per cent of Zimbabwe's total export income, and in Malawi an even greater proportion, 68 per cent, comes from the export of tobacco. Tobacco production is completely dominated by large companies, both national monopolies and multinational enterprises. Seven companies account for half of the global production.

The negative effects of tobacco *consumption* also increasingly affect the developing world. Tobacco is one of the most profitable consumer products, but the market has shrunk significantly in the industrialized world. Cigarette marketing is increasingly restricted, and fewer people smoke. Multinational tobacco companies therefore market themselves in developing countries, although they cannot charge such high prices there. The largest producer in the USA, Philip Morris, sold more cigarettes outside the USA than domestically for the first time in 1990. By the year 2005, the greatest profits are also expected to be earned outside the USA.

Sources: French, H F (1993) 'Costly Tradeoffs. Reconciling trade and the environment' Worldwatch Institute Paper 113, p.12; Wells, P and Jetter, M (1991) *The Global Consumer* Victor Gollancz, London p.180. 'The search for El Dorado, The Tobacco Trade' *The Economist* 16 May 1992, p.21 f: 'Hard to give up the weed' *The Economist* January 1986, p.56.

- Consumption of imported goods causes local environmental damage.
- Production causes local environmental damage, and the country producing goods does not take environmental costs into account.
- Pollution travels across borders as a result of consumption or production.

In the first case, damage is connected to products and is a direct result of their consumption. In the second case, which may be referred to as environmental

Box 7
THE POLLUTER PAYS

In environmental economic theory, it is sometimes stated that either the polluter pays or that the person or persons who suffer pollution pay ('the polluter pays' or 'the victim pays'). The determining factor is who has the ownership rights to the environment, or who has the right to demand compensation from whom.

In this example, A is the polluter and B is the victim. A victim who has complete rights to a clean environment can demand that all polluting activities are terminated. The last marginal purification is expensive for the polluter, however, while the perceived improvement to B is slight. Both profit when A purifies to the point where the cost of purification is the same as the perceived advantage, and thereafter A compensates B financially so B is as satisfied with the situation as if there was no pollution (Figure A).

The supply curve (person A) shows the cost of purification while the demand curve (person B) shows the perceived benefit of purification. The optimal level of purification is indicated by a star (*).

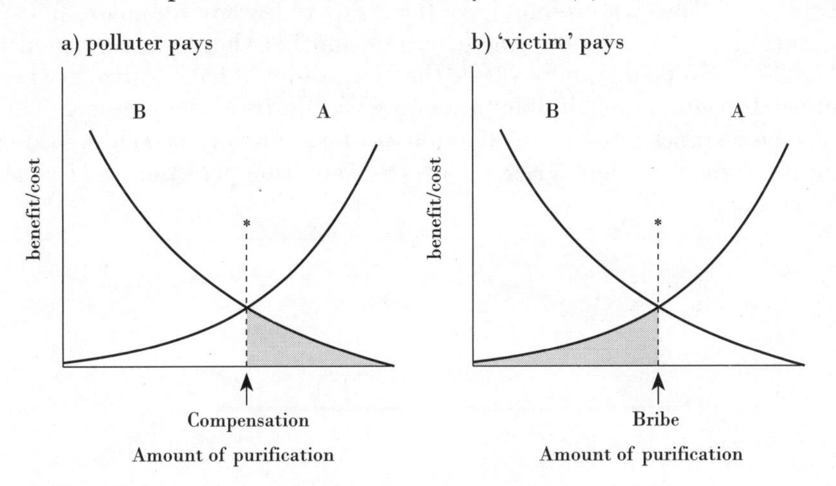

If the polluter instead has the right to pollute the environment, then we have a situation where the victim pays. If B desires the emissions to be cleaned he must pay (bribe) A in order for those measures to be carried out. A will pay as long as he perceives a certain satisfaction and is not troubled by the cost of the measures (Figure B).

The advantage of this reasoning is that, regardless of how ownership rights are decided, the optimal anti-pollution measures will be adopted in both cases. This piece of economic theory should not be confused with PPP – the Polluter Pays Principle – as adopted by the OECD in 1972 and later in Rio in 1992. PPP is a principle for how anti-pollution costs and the costs for avoiding environmental damage should be distributed. The principle requires companies to bear the costs of introducing cleaning technologies and not receive subsidies from the government. It does not give any direction as to how costs should be distributed for damage which occurs despite anti-pollution measures, and does not suggest any particular system for compensation. Neither does it state the level to which emissions should be abated. Since 1975, it has also been permitted to demand compensation, but due to strong opposition from industry no country has actually utilized this tougher form of PPP.

An important reason for the OECD, at an early stage, to establish principles was the wish to avoid a distortion of world market prices. If cleaning costs are distributed differently in different countries, so that one country subsidizes a company while another lets the producers pay, the price of goods would be affected.

From the viewpoint of environmental economic reasoning, we see that PPP does not give the polluter complete responsibility for paying, since it indicates how the cost of *purification* can be distributed but does not present a clear picture of how the cost of *environmental damage* should be paid.

It is not obvious who should have the rights to the environment, and it is apparently not the case that one group or another should have unlimited rights. There can be instances where the victim should have a stronger right – and would thereby be eligible for compensation from the polluter. There can also be instances where effective purification can only be achieved if the victim pays, such as when a rich country suffers from pollution occurring in a poor one.

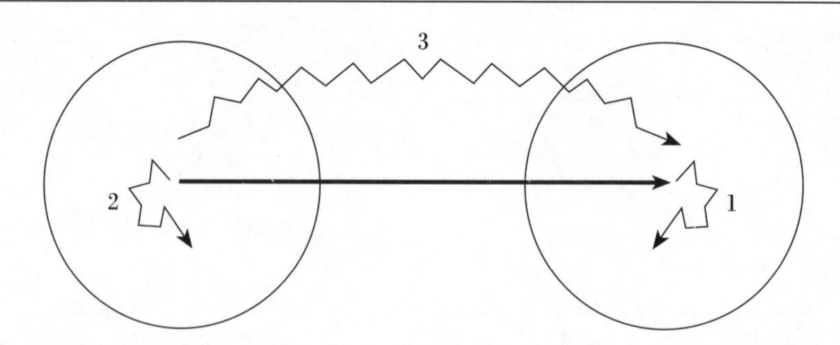

Figure 15 Three fundamental effects with one country causing damage to another due to the environment's transnational dimensions. The straight line depicts trade flow. The crooked lines represent environmental effects.

dumping, the problem stems instead from production processes. Damage is an indirect result of costs to implement domestic environmental policy since it may cause production to move abroad. In the third case, which again may be associated with production, a country can be damaged directly by transfer of pollution as well as indirectly because environmental action is counteracted, since production may relocate and continue polluting abroad. In all three situations, there are problems in adopting environmental legislation due to activities in another country.

Most international environmental issues can be classified in accordance with these three effects. Some issues will display elements of more than one effect (see also the discussion on contamination due to waste disposal in Box 5). Of course, there may be more than two countries involved. The transportation of goods commonly causes environmental damage in countries other than those that export or import (see Chapter 1). In the case of global systems common to all countries, such as the sea and the atmosphere, everyone will eventually suffer from mismanagement. For simplification, however, we will take examples from the two countries model.

In the following section, we consider what role harmonization, trade barriers and other policy tools may play in the three cases of environmental damage.

Should environmental norms be harmonized?

Because of the environment's international dimensions, individual countries cannot usually push through measures which make producers pay the actual cost of environmental damage. Uncoordinated actions in different countries lead to a result which is undesirable for all parties.[89] Perhaps the most commonly heard suggestion for the management of international environmental problems is that emission levels and norms should be harmonized, that is standardized across countries. The problem is that harmonization requires similar environmental concern from each party. If the same demands were made in all countries, there would be no need for coordinating policies.

The fact is, however, that the environment, together with capital, labour, technology, and so on, forms the natural conditions for production. Adjustments are constantly made – whether consciously or not – between the environment and commercial values, which decide the level of ambition for policy. Adjustments are affected by several factors, which vary between countries. These factors include:

- Natural conditions differ. One example is the ability of ecosystems to absorb emission; it is more damaging to emit a kilo of sulphur over an acidic forest in Sweden's Bohus province than over calcium-rich land in Ireland.
- The costs for alleviating environmentally damaging activities differ from country to country; for example, it is less expensive to purify a kilo of sulphur in Poland than in Sweden.
- People's incomes and willingness to pay for a 'clean environment' vary both from country to country and within a country. The occurrence of

89 For early observations on this matter see Ward, B and Dubos, R (1972) *Only One Earth* Penguin; Dasgupta, B (1976) *Environment and Development* UNDP, Nairobi

ulations of different countries, hampering competition and economies of scale, especially in small markets. For such reasons, efforts are being made, for example within the European Union, to achieve common standards.

It is far from obvious on which level product norms should be decided – local, regional, national or international. Global standardization of environmental requirements is generally not a good idea, while decisions made on excessively local levels lead to losses in efficiency, and possibly to abuse. The fact remains, however, that product standards must be introduced at some level, as they constitute the accurate means of managing the environmental effects which result from consumption, both when the goods are produced domestically and when they are imported.

Effects of production processes

The remaining two effects by which one country suffers because of production activities in other countries are more difficult to rectify with trade barriers. It should be noted that the acceptance of unregulated environmental effects is synonymous with the implicit subsidy of polluting activities, at the cost of other countries when damage becomes transnational.

Trade barriers may be justified when environmental damage occurs as a result of manufacturing processes in other countries. The desire for trade barriers comes from a lack of alternative measures which would directly address the source of the problem, and the belief that trade barriers would induce the producing country to change its actions. The desirable level of duty, or other trade measure, depends on the damage suffered by the importing country and how trade measures – which are directed towards products – influence the conditions for the manufacturing processes which represent the underlying cause. As shown by Baumol (1971), the desirable level of duties may vary depending on whether the perspective is that of an individual country or that of all countries.[94]

Trade barriers may, however, have unexpected effects, such as an undesired distortion of prices or triggering of retaliatory action.[95] Furthermore, it may be virtually impossible for small and economically weak countries, which lack negotiating power, to apply them. Finally, it should be noted that other factors speak against trade barriers as a means of addressing environmental issues. Although these factors are primarily associated with the preconditions for policy-making, they need to be included in an economic analysis.

Trade with or without internalization

As long as environmental costs are reflected in market prices, we have stated that trade does not, itself, cause environmental problems. Under such circumstances, a liberalization of world trade would be compatible with, and would

94 Baumol, W (1971) *Environmental Protection, International Spillovers and Trade* Almqvist and Wiksell, Stockholm

95 Anderson, K and Blackhurst, R (eds.) (1992) *The Greening of World Trade Issues* Harvester Wheatsheaf, New York

support, long-term sustainable development. We have, however, seen that in general environmental costs are not internalized. This raises the following two central questions:

- Does present development lead to an increased or a reduced internalization of environmental costs?
- Is liberalization of world trade compatible with ecologically sustainable development when environmental costs are not internalized?

There is little doubt that the public is more aware of environmental issues today than in previous decades and that substantial investments are being made in many areas to alleviate environmental problems. With the signing of the Rio Declaration and Agenda 21, most countries have at least in principle agreed to strive for internalization of environmental costs. The significance of this position is, however, highly unclear. Principle 16 of the Rio Declaration states for example that 'National authorities should endeavour to promote the internalization of environmental costs and the use of economic instruments . . . without distorting international trade and investment.' (see Appendix 2 for the full quotation.)

In fact, there is no doubt that in many cases an internalization of environmental costs would lead to significant changes in production and consumption patterns, which would in turn affect trade as it is today. Products which burden the environment would become less common and be replaced by those which are more resource-effective and ecocycle-adjustable. Reality indicates that the process towards an internalization of environmental costs is still in its infancy. Certain countries have come further than others, but, overall, a great deal remains to be done. Very few countries apply economic policies (such as taxes and fines) to counter environmental degradation, and legislation is, in many cases, insufficient. This has been discussed in the previous chapter.

The answer to the question of whether increased liberalization of world trade is compatible with the goal of long-term sustainable development depends on the extent and speed with which environmental effects are internalized. If environmental effects are not heeded, it is very likely that an expanding world economy will become incompatible with the life-supporting ecosystems of our planet. It is, therefore, far from certain whether liberalization of trade is in the common interest of humanity.

An expansion of trade may increase material welfare, and lead to higher demands for environmental measures, as well as creating more financial resources and better technology for the internalization of environmental costs. But, for this to be realized, the market players must receive signals that cause them to believe in the profitability of these measures. Furthermore, the environmental effects of the trade expansion itself must also be internalized. For companies and households to avoid being threatened by costly adjustments when the environmental problems which have long been neglected suddenly become acute and 'force' measures, mechanisms are needed now which trigger a long-term goal formulation of structural conversion. This boils down to creating institutional relationships and frameworks which effectively stimulate sustainable production and consumption patterns.

LOCALIZATION OF ENVIRONMENTALLY DAMAGING ACTIVITIES

Through the occurrence of trade and deregulation of factor markets, it is theoretically possible to localize environmentally damaging activities where they do the least damage and/or where there is no requirement to take environmental destruction into account. The likelihood that production can be knocked out or relocated to another country may, at the same time, prevent demands for long-term environmental consideration in one country when corresponding measures are not introduced by other countries. In this section we discuss the questions surrounding the relocation of polluting production activities.

Expected relocation

It was predicted as early as the 1970s that long-term anti-pollution requirements in certain countries would lead to the relocation of polluting activities – in practice from developed to developing countries. Multinational firms were believed to play an important role in the process, which was greater the less dependent companies were on country-specific input goods.[96] Little comprehensive relocation has been observed, however, with the exception of production of highly poisonous substances such as asbestos, benzidine colouring agents, chemical pesticides and heavy metals like copper, zinc and lead. In these cases, activities have been relocated to developing countries after the closure of plants in developed countries. According to Leonard,[97] relocation has only taken place in industries with slow growth and weak domestic demand and which have not significantly contributed to economic development.

Although the developing world serves as a base for a relatively large share of environmentally damaging industry, there has been considerable speculation as to why a greater degree of relocation has not been observed.[98] One position is that the costs of pollution abatement are too small to influence localization. Achieving the technological advancement needed to fulfil these demands can also be preferable to relocation for a company. Further, there is a risk that other countries will also propose similiar requirements in the future and that con-

96 Walter, I (1972) 'Environmental Control and Patterns of International Trade and Investment: An Emerging Policy Issue' Banca Nazionale Del Lavoro Quarterly Review, 100, pp 82–106; Pearson, C (1976) 'Implications for the Trade and Investment of Developing Countries of United States Environmental Controls' United Nations Conference on Trade and Development, Geneva

97 Leonard, JH (1988) *Pollution and the Struggle for World Product* Cambridge University Press

98 Low, P and Yeats, A (1992) 'Do Dirty Industries Migrate?' In Low, P (ed.) 'International Trade and the Environment' World Bank Discussion Paper no. 159, Washington, pp 89–104; Lucas, REB, Wheeler, D, Hettige, H (1992) 'Economic Development, Environmental Regulation and the International Migration of Toxic Industrial Pollution: 1960–1988'. In Low, P ibid, pp. 67–86

sumers will react negatively to a company which moves its pollution-producing activities abroad rather than try to counteract the pollution.[99] In an unofficial memo from the World Bank in 1992, it was asserted that more relocation is desirable and should be encouraged (Box 5).

Is relocation desirable?

The supposition that relocation as a response to strict environmental policies poses no problem, does not necessarily mean that production should move to the location where environmental protection is the weakest, however. It is relevant to discuss briefly the arguments for and against localization of environmentally destructive activities in developing countries, which tend to have few protective measures.

It is often asserted that ecological consideration is a luxury which developing countries cannot afford today, and that acute problems such as hunger, illiteracy and unemployment should take precedence over conservation of the environment. In other words, the socially desirable level of protection for the environment would be lower than in industrialized countries. This argument may be countered because developing countries have an especially limited ability to manage environmental problems. Emissions are often associated with risk-taking, and developing countries have limited resources to manage a negative outcome. Many biological and chemical substances spread much more easily in a tropical climate, even though some elements might also break down more rapidly. A population suffering from chronic malnutrition has a reduced chance of fighting off illnesses. Furthermore, industries in developing countries are generally geographically concentrated and efforts to spread them out have mostly failed. While it was possible, some decades ago, to assert that developing countries had a relatively undisturbed environment the case can rarely be made today.

General relocation of environmentally damaging activities to developing countries is therefore undesirable, even though it may be justified in some cases. The evidence that a limited number of relocations have taken place might suggest that the potential profits are not sufficient to cover the costs involved in the move. It is probable, however, that the possibility of relocation does affect policy in many countries.

The absence of internalization

The relocation of polluting activities may have negative consequences if not all countries apply measures which force the internalization of environmental

99 Gladwin, TN and Welles, JG (1976) 'Environmental Policy and Multinational Corporate Strategy'. In Walter, I (ed.) *Studies in International Environmental Economics* John Wiley, New York; Leonard, HJ (1984) *Are Environmental Regulations Driving United States Industry Overseas? An Issue Report* The Conservation Foundation, Washington; Pearson, C and Pryor, A (1978) *Environment North and South: An Economic Interpretation* Wiley-Interscience, New York; Tobey JA (1990) 'The Effects of Domestic Environmental Policies on Patterns of World Trade; An Empirical Test' *Kyklos* 43, pp. 191–209

effects. Given that the level of environmental protection is 'excessively low' in some countries, activities may be relocated even though this will lead to a less productive allocation of resources. The risk of such 'incorrect' relocations can motivate trade policy measures as a complement to environmental measures – given that industries are thereby prevented from moving abroad, continuing to pollute and export from the new country.

Other measures are generally more effective than trade barriers at preventing the relocation of production. One policy option is to subsidize anti-pollution measures, which is often touted as the 'next best' instrument for managing environmental effects (after a tax which equates marginal costs of emission and purification).[100] Subsidies are also associated with problems, however. The individual company's emissions are reduced, but entry of new companies may be stimulated, so that there is risk of a *total* increase in emissions.[101] In order to avoid this, it is essential that subsidies are designed so that technical development of environmentally improving measures is stimulated.[102] Depending on their form, subsidies may also violate existing trade agreements.

Technological transfers

Relocation of production through multinational corporations involves technological transfers, which also spill over to local players in host countries through the exchange of labour, instructions to suppliers, contacts with local authorities, and so on. Indeed, multinationals today play a leading role in the world economy with respect to both development and international diffusion of new technologies. It is important to differentiate between the availability of existing advanced environmental technology and how it is actually used, however. The most advanced technology is not generally demanded by subsidiary companies in developing countries and it is seldom forced upon them by the parent company. Multinational companies tend, however, to have cleaner production in developing countries than domestic companies, even if it is more environmentally damaging than the equivalent production in industrialized countries.[103]

If foreign companies' investments replace or influence domestically-owned production, there is a large potential for improved environmental management in poorer countries. This does not imply that foreign technology is always preferable to the technology of the host country. Some studies have pointed to a generally low level of technological adaptation by multinationals to the special conditions present in developing countries.

To what extent multinational companies transfer environmentally-adapted

100 Mestelman, S (1982) 'Production Externalities and Corrective Subsidies: A General Equilibrium Analysis' *Journal of Environmental Economics and Management*, June, pp. 186–93

101 Baumol, WJ and Oates, W (1988) *The Theory of Environmental Policy* second edition, Cambridge University Press

102 Harford, JD (1991) Pollution Control Cost Subsidies and the Enforcement of Standards, mimeo, Cleveland State University

103 United Nations Centre on Transnational Corporations (UNCTC) (1985) *Environmental Aspects of the Activities of Transnational Corporations: A Survey* New York

technology or use an absence of cleaning requirements in order to avoid taking action over emissions is greatly influenced by the actions of consumers. Today, environmentally-friendly businesses and products are more and more appreciated, which makes it possible for a company to strengthen its good reputation by taking global responsibility (see Box 1).

> *What I would ultimately like to be able to do is set up a perfect example of honest trading in a fragile community and make it a benchmark of how we should conduct such trade in future. The rules are pretty simple. First, we have to be invited. Second, we must not mess with the environment or the culture. Third, we must reward the primary producers.*
>
> Roddick, A (founder of The Body Shop) (1991) *Body and Soul* p.213.

The more reliable and available information is on a company's actions, and the better the consumer can evaluate them, the more motivated companies will be to consider the environment.[104] Those who have not been successful in this area are motivated to obstruct factual information, while others have reasons to encourage it. Opposing interests similarly prevail between governments, partly due to differences in ability between their respective industries. Countries around the world have negotiated in vain for years to establish a 'code of conduct' for the actions of multinational companies with regard to environmental as well as other issues.

POLITICAL FACTORS

As stated earlier, it is difficult to manage the international environmental problems which result from environmental dumping or unregulated transboundary effects. There are reasons to probe more deeply into the real cause of the problem. The fact remains that environmental dumping is generally inferior to other methods as an instrument for increasing competitiveness.[105] Further, it might be expected that countries would be capable of managing international environmental issues either through binding commitments and/or through the provision of transfer payments. That this does not always occur highlights the motives of those in power and how these affect the interaction between players. It has traditionally been assumed that governments act in the interests of their citizens. The 'public choice' school has however pointed out that it cannot be assumed that policy makers act to maximize society's good. The opposite seems to be true: some people in power may look to their own profit, which causes the inter-

104 Flaherty, M and Rappaport, A (1991) *Multinational Corporations and the Environment: A Survey of Global Practices* Center for Environmental Management, Tufts University, Boston; Gladwin, TN (1977) *Environment, Planning and the Multinational Corporation* Jai Press, Greenwich

105 Barrett, S (1993) Environmental Standards, Imperfect Competition and International Trade, mimeo, London Business School

ests of influential groups to weigh relatively heavily. The possibility of politicians seeking to maximize something other than society's welfare demands that the functioning of public institutions be included in the overall economic analysis. This is also underlined by the fact that so many of the world's countries are governed by dictatorships.

Skewed political influence

The formulation of trade policies is influenced by society's power structure. The benefits of a healthy environment tend to be relatively long-term and spread out over many people, while the costs of combating problems are relatively short-term and concentrated on smaller groups. Those who will eventually suffer from damage are perhaps not even born when it is caused and cannot speak for themselves at that time.

Again, it should be stressed that complete information is rarely available and the different players do not have access to the same information. Environmental effects can be made less politically sensitive if they are channelled against groups who are uninformed about, or have limited opportunities to protest against, their losses. For example, pharmaceuticals, poisons and other goods which cause dangerous health effects may be dumped in countries and areas where either the authorities or the public have particularly small means to evaluate them. One example is the ongoing redistribution of tobacco use from wealthy to poor countries, which is supported by comprehensive advertising campaigns (see Box 6). Another aspect of political influence is that certain spectacular environmental issues may be exploited as an excuse for measures which fulfil completely different objectives.

Against this background, there are likely to be distortions in political influence which discriminate against the environment. At the same time, the introduction of trade barriers benefits well-organized producers while harming consumers.[106] Anti-dumping policies in the USA and the EU present disturbing examples.[107] For such reasons, trade barriers and consideration for the environment can be said to represent an unholy alliance. Differently stated, it may be difficult for those concerned with the environment to assert themselves politically against those who demand trade barriers for other reasons.[108] There is a definite need for general regulations which cannot be manipulated and evaded, if the poor are to be protected. This is demonstrated clearly by the fact that the two types of production within which developing countries have clear-cut comparative advantages – agriculture and textile manufacturing – have largely landed outside the GATT rules and are characterized by a tangle of trade barriers and subsidies.

106 Olson, M (1965) *The Logic of Collective Action* Harvard University Press, Cambridge, Mass.
107 Hillman, AL and Ursprung, HW (1992) 'The Political Economy of Interactions between Environmental and Trade Policies' In Anderson, K and Blackhurst, R (eds.) *The Greening of World Trade Issues* Harvester Wheatsheaf, New York, pp. 195–220
108 Finger, M (1993) World Bank, Washington, in press

Box 8
RAIN FORESTS AND INTERNATIONAL TRADE

Nearly half of the world's rain forests have disappeared. During the 1980s, destruction of the forests increased by 50 per cent, and 17 million hectares of tropical forests are cut down annually: an area equal to 50 per cent the size of Germany. The rate of felling is rising most rapidly in Central Africa, the Caribbean and South-East Asia. Logging is estimated to account for 30 per cent of the increase in atmospheric carbon dioxide and 20 per cent of the total greenhouse effect.

Several factors contribute to the processes which destroy the forests. Deforested areas are nearly always converted into agricultural areas, either as large plantations of, for example oil palms or rubber trees, or more often small-scale cultivation by poor farmers. They generally move to the rain forest because they have been forced off their own land. Agriculture is not however the main cause of deforestation; the forest has, in most cases, already been opened up by loggers. Logging in the rain forest does not have to mean environmental degradation but it must be managed in a sustainable manner.

Table 6 Tropical deforestation

Forest areas and annual deforestation in tropical forests, the total for each section of the world and examples from individual countries and smaller regions.	Shadow areas 1980, per 1000 hectares	Annual felling per 1000 hectares	Annual felling %
Latin America	923,000	8300	0.9
Brazil	500,000	2500	0.8
Asia	310,800	3600	1.2
Malaysia	113,900	1000	1.5
Indonesia	21,000	270	1.0
Africa	650,300	5000	0.8
West Africa	55,200	1200	2.1
Central Africa	230,100	1500	0.6
Total	1,884,100	16,900	0.9

Source: *World Resources 1992–93*, reworking of Tables 8.2 and 19.2. The figures are uncertain and are based upon assessments which are several years old.

Development projects such as road building, power station dams and mining also destroy forests and open the way for small farmers. In Latin America, large tracts of forest have been converted into pasture for beef cattle, often after first being occupied (and then abandoned) by farmers with small-holdings. In 1988–89 in Brazil, tax regulations which directly encouraged clear felling of the forest were altered, but a failed agricultural reform meant that people seeking land still move to the Amazon area.

Timber trade and the rain forest in South-East Asia

South-East Asia is a region where the link between commercial felling and the destruction of the forests is most apparent. The region provides 87 per cent of the total export of tropical timber. Malaysia alone provides half of the total. The greatest proportion is exported to Japan, where it is used for moulds in the construction business.

Only a small proportion of the value of the forest is reflected in timber prices. Some examples of actual, but difficult to measure, value are the climate-regulating and soil-conserving functions of the forest, gene resources and the commercial value of items other than trees. Protests against the destruction of the rain forests have grown during recent years.

Restrict imports?

In the western world growing concern for the consequences of deforestation has led to a demand for decreased importation of tropical timber. Germany stopped the use of tropical timber in 1989 and the USA has prohibited imports from Myanmar (Burma). In 1990, the European Parliament adopted a resolution that the EU should regulate the import of tropical timber and also create a fund to support the development and realization of plans for sustainable forestry in tropical countries. According to the proposal, licenses will be required for the import of tropical timber. The licenses will be based on annual negotiations with export countries and connected to the realization of forest conservation plans.

In 1992, Austria introduced a law on obligatory labelling of goods made from or containing tropical timber. At the same time, quality marking of goods was introduced, with high requirements for products which come from 'enduring forestry'. Singapore, Brazil, Malaysia and other timber exporting countries questioned the laws and threatened to restrict all their imports from Austria. The countries felt that the labelling requirements were discriminatory, since timber from temperate climates did not need to be marked, and that the marking would give a negative image even if the import of timber was not formally prevented. Timber exporters also protested because Austria had reacted on her own in defining what could be regarded as sustainable forestry. On 1 April 1993 the mandatory labelling of goods with their place of origin was discarded.

Generally, it can be said that reduced exports lead to lower prices, but opinions are divided on what effect price changes will have on rain forest timber. Many observers believe that lower timber prices would make the forest less profitable in relation to other ways of using the land, so that people would have even greater reason to convert the forest into agricultural land. A ban on imports could therefore lead to increased deforestation.

Other experts believe that if ownership rights are poorly defined, which they generally are, then higher timber prices would lead to more rapid felling. If a felling company has permission to cut down trees for a specific period and does not know if it will gain from the future profits from the forest, that company will be quick to fell trees while prices are high and the opportunity is available. Many people also believe that since felling nearly always opens up the forest for agriculture, it is a necessary prerequisite for nearly all conversion of forested areas. Reduced felling for export is therefore of vital strategic importance.

Export ban for raw timber

Indonesia has successively raised tariffs on the export of raw timber and in 1984 introduced a total export ban. The explanation is that by retaining a larger proportion of the value added within the country they will not need to fell trees as quickly in order to earn the same income. The export ban has resulted in a rapid increase of plywood and veneer manufacturing which have completely replaced timber export. The export of timber products today accounts for 14 per cent of the country's total exports, and the industry employs 3.7 million people directly and 15 million indirectly.

The export ban points out the importance of many of the trade barriers which the Western world has applied, compared to developing countries, and which makes it difficult for them to develop refining industries. For example, customs duties for timber products have risen at the same tempo as the rate of refining in many countries. The Indonesian export ban can therefore be said to be the answer to the tariffs which, for example, Japan has imposed on refined wood.

The export ban is frequently discussed. Some people feel that an export ban will, in the long run, result in a reduction of logging even though it may not seem to have had that effect so far. A local refining industry has more reason to conserve the forest than a felling company which can move its activities to another country when the forest resources have been totally used up. Other observers believe that domestic industry will intensify deforestation, since refining is less effective than in the western world. The export ban is being questioned by, among others, the EU, and a GATT panel is discussing the issue.

Other possibilities to arrest the plundering of the rain forest

As with many other questions where both trade and the environment are involved, it is not trade itself which causes the problem. The root of the problem is often internal policy in individual countries. Trade policy is therefore not the best solution, even though in certain cases it seems to be one of the only practical possibilities, and may also result in a number of undesirable side effects. What other possibilities are there to influence the internal policies of tropical countries when the sovereign rights of a country must be respected according to international law?

- Firstly, nothing prevents organizations and individual consumers from halting the purchase of tropical timber felled using non-sustainable practices. One example of this is that Swedish architects have halted the use of rain forest timber.
- The initiative has been taken to form an organization with the aim of developing global, environmental labelling of products originating from tropical as well as temperate forests. Environmental organizations, timber companies, governmental authorities, environmental labelling organizations and local populations will be represented in the Forest Stewardship Council (FSC). Ten general principles for intelligent forestry, complemented by local and more detailed criteria, will be the basis for labelling.
- The few existing examples of sustainable utilization of the rainforest offer an alternative to clear cutting. A newly formed British company, the Ecological Trading Company, works cooperatively with the local population and imports timber felled in a socially and ecologically acceptable manner.
- By increasing information and supporting the positive forces in tropical countries, their policies can be influenced. These forces can, for example, be individual organizations, organizations of local populations or governmental authorities.

The most important thing is to influence the fundamental reasons for deforestation, otherwise a ban on logging, for example, will have no effect. A total ban on felling was introduced in Thailand, but illegal felling continued and many timber companies moved to Myanmar (Burma). Illegal felling and trade of timber products occurs on a massive scale. In South-East Asia alone, hundreds of thousands of hectares of land are illegally felled or exported every year.

A global problem

Since the diversity of species contained in the rain forests can be considered to be our common inheritance, and since the deforestation of rain forests contributes to the greenhouse effect, deforestation in tropical forests must be regarded as an international problem. Does this mean that we should no longer accept the choices made by a country's government, but instead take the matter into our own hands by introducing trade barriers when deforestation so obviously conflicts with long-term sustainable development?

It is important to find cooperative methods which take all the parties' interests into account. Perhaps the most important condition needed to stop deforestation is that people who live in and near the forest areas be allowed to enjoy a part of their real value. International measures to protect tropical forests must therefore contain some form of compensation to rain forest countries for, among other things, gene resources and carbon dioxide absorption services. Developing countries' governments today require that all forests, not just rain forests, are included in the international cooperation on forest protection. Protection of the biodiversity in Swedish forests may prove to be essential to the saving of the rain forests.

Sources:
Alberto, C and Braga, P (1992) 'Tropical Forests and Trade Policy: The Case of Indonesia and Brazil' World Bank Discussion Papers 159
Barbier, E B and Rauscher, M (1992) 'Trade, Tropical Deforestation and Policy Interventions' Beijer Discussion Paper Series no 15
Burgess, J C (1991) 'Timber Production, Timber Trade and Tropical Deforestation' *Ambio* 20:2–8
Callister, D J (1992) 'Illegal Tropical Timber Trade: Asia-Pacific' TRAFFIC Network
Daly, H E and Goodland, R (1994) 'An Ecological-Economic Assessment of Deregulation of International Commerce under GATT' *Ecological Economics* 9:73–92
Myers, N (1989) *Deforestation rates in tropical forests and their climatic implications*, Friends of the Earth
Nectoux, F amd Kuroda, Y (1990) *Timber from the South Seas*. WWF International
The Forest Stewardship Council, A discussion paper, 1992
World Resources 92–93, page 120
Wright, M 'Selling timber without selling out', *Tomorrow*, page 87 no. 2, 1991.

Despite broad-based support for the principles of free trade, the global community had great problems in completing the latest round of Uruguay discussions in GATT. The future for open world trade remains uncertain. Inasmuch as it is not possible to clarify with great precision when trade barriers are used to protect the environment and for no other reason, there is an obvious risk that the environmental argument can be used as an excuse for traditional protectionism. With the door open to discriminating trade policies, the prerequisites for cooperation on common resource questions deteriorate even further.

The interaction between interested parties

For policy-making to produce wanted effects, it is necessary to consider how people adapt their actions, and the basis of how they expect others to act. The interaction between different interested parties is expressed in strategic behaviour. A well-known example is the 'prisoner's dilemma' – two people who cannot coordinate their actions both manage to end up in trouble, since each one risks being implicated by the other. Another example is when one party tries to hitch a free ride from another.

Combined with imbalances in political influence, strategic behaviour leads to extremely complex decision-making. Even if the ruling power in one country strives for society's greatest good, it may have to cooperate with other countries whose governments are indifferent to the state of the environment. In effect, the failure to heed environmental consequences in one country may reduce the level of desirable environmental requirements in other countries.[109]

In the case of localization of production, countries may outbid each other with incentives to attract investors.[110] Companies, on the other hand, can be expected to make use of this situation for their own advantage. One example is the gradual devastation of natural resources in one country after another, which has been referred to as 'sequential exploitation'.[111] The fundamental reason behind such processes is the inability or disinterest of host countries in coordinating policies between them. Poorly functioning capital markets, exacerbated by the debt crisis, contribute to the problem by limiting access to capital, raising interest rates and creating extremely limited planning horizons for economic decisions.

There are several possible explanations as to why environmentally damaging activities are not relocated more often. If the majority of countries set low priorities on environmental protection, then each country may be forced to be satisfied with particularly low environmental demands in order to avoid making industries move. If other countries do not care about the environment, those who do care may consequently be lured to unwanted action. This may include industrial policies, subsidies, requirements for trade barriers or too low environmental demands. The potential relocation of environmentally damaging operations across geographical borders therefore impacts on the desire and ability to protect the environment, even though a more comprehensive relocation is not undertaken.

Because of asymmetries in the distribution of information, the limited representation of many people in the political process, and the fact that coordination of all interests cannot occur, it is impossible to come up with ideal solutions for many international environmental issues. Feasible progress requires that the interaction between various interests is used positively, resulting in a fruitful upgrading of real demands among the players involved on different levels – government authorities, companies and consumers around the world.[112]

An example of cooperation

Environmental charges made by a country result in companies incurring costs

109 Andersson, T (1991) 'Government Failure – the Cause of Global Environmental Mismanagement' *Ecological Economics* 4 pp. 215–36

110 Markusen, JR, Morey, ER and Olewiler, N (1993) 'Noncooperative Equilibria in Regional Environmental Policies when Plant Locations are Endogeneous' NBER Working Paper 4051

111 Berkes, F (1985) 'The Common Property Resource Problem and the Creation of Limited Property Rights' *Human Ecology* 13:187–208

112 Cooper, R and John, A (1988) 'Coordinating Coordination Failures in Keynesian Models' *Quarterly Journal of Economics*, CIII, pp. 441–63

in the short-run. On the other hand, technological development and skills which reduce costs are stimulated. Given that companies win 'goodwill' in the eyes of the consumer and that environmentally damaging activities are met by increasing international opposition, other countries are pressured eventually to follow their lead. Those who introduce strict environmental legislation at an early stage enjoy a strategic advantage, which has so far been exploited by, for example, Japan and Germany. Porter, an American professor, argues that the US should take the lead.[113]

> *The World is moving toward deregulation, private initiatives, and global markets. This requires corporations to assume more social, economic, and environmental responsibility in defining their roles. We must expand our concept of those who have a stake in our operations to include not only employees and shareholders but also suppliers, customers, neighbors, citizens' groups, and others. Appropriate communication with these stakeholders will help us to refine continually our visions, strategies, and actions.*
>
> Declaration of the Business Council for Sustainable Development (1992)

If consumers do not appreciate anti-pollution efforts, and other countries are not expected to adopt them, then companies can instead respond to environmental demands in one country by relocating to another in order to avoid the charges. How consumers and companies react in turn decides whether other governments find it worthwhile to require environmental protection as well – since their reaction decides to what extent they attract investors or lag behind in the technological race if they do not.

The combination of interactions, illustrated in Figure 16 for the case of purification requirements, decides how fruitful it is for an individual country to take action in the first place.

In order for transboundary environmental policies to be effective, it is essential that international measures are coordinated. For example, governments must all aim to reduce the risk of accelerated greenhouse effects as a result of increased carbon dioxide in the atmosphere. Calculations which heed technological advancements show how sensitive a national economy is to the pace at which a tax on carbon dioxide is introduced. Measures – if shown to be necessary – will be more costly the longer implementation is delayed.[114]

Effective systems of charges must be created internationally and, if at all possible, implemented internationally to have the potential to achieve the most efficient reduction of pollution. On the other hand, the desire to adopt measures increases with the ability to pay. The OECD has calculated costs for possible strategies. All developed countries, or OECD Europe, could introduce

113 Porter, ME (1990) *Competitive Advantage of Nations* Free Press, New York

114 Jorgenson, DW and Wilcoxen, P (1993) 'Reducing US Carbon Emissions: An Econometric General Equilibrium Assessment' *Resource and Energy Economics* 15. no1

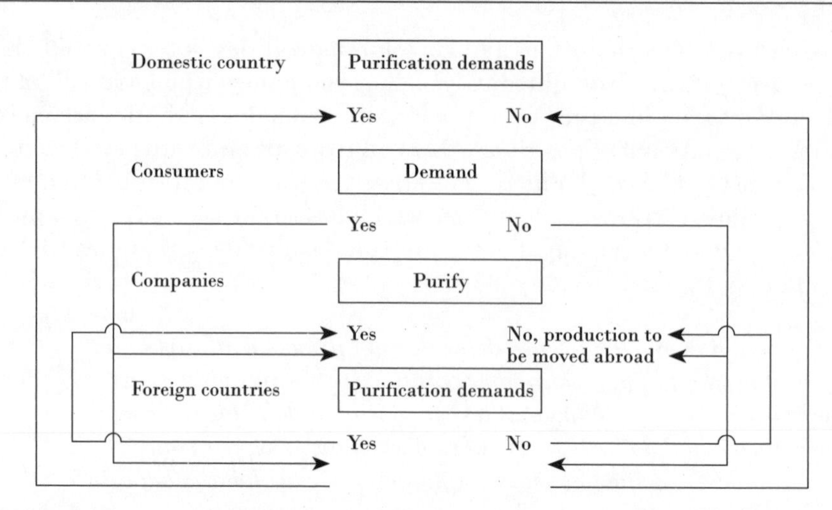

Figure 16 Interactions among consumers, companies and governments. Each player's judgements and actions determine the outcome.

reasonable charges for the emission of carbon dioxide without suffering unmanageable costs.[115] Although the cost of reducing emissions is lower in poor countries, measures to achieve this will not be implemented there unless industrialized countries introduce them first. If the OECD adopts modest but gradually rising charges for emissions, technological advancement will be promoted, and at the same time the means will be created to introduce the measures in poorer countries.

A superior solution would be achieved by coordinating measures which influence both emissions and absorption of carbon dioxide. For example, a more efficient way to reduce human contribution to the greenhouse effect would be to check forest destruction rather than limiting consumption of fossil fuel. Besides the carbon dioxide that is released when forests such as those in the Amazon area are burnt, the destruction of tropical rain forests limits the potentially valuable genetic diversity. The losses, for example in the form of unexploited opportunities for medicines in the future, increase more when ongoing advances in biological technology open new areas of application for genetic information.

Developing countries, however, lack the capital and technology which would make it possible for them to profit from the wealth of the rain forests. While access to credit is rationed, previously accumulated debt burdens and high interest rates generate a great need for capital income in the short term. The establishment of industry is also hindered by trade barriers on processed products in developed countries. Developing countries therefore have little sympathy for demands, by the developed world, that the rain forest be preserved (Box 8). There are, naturally, many other ecosystems which are

115 OECD (1992) 'The Economic Costs of Reducing CO_2 Emissions' *OECD Economic Studies*, no.19/Winter, Paris

important to manage with care, for example sensitive mountain and coastal areas. A great number of ecosystems are under serious threat in the 1990s and, like most of the tropical rain forests which still remain, will disappear within a few generations, unless a satisfactory share of the actual value of their resources goes to the people living with, and from, them.

POSSIBLE SOLUTIONS TO INTERNATIONAL ENVIRONMENTAL ISSUES

This chapter has discussed the economic view of the connection between trade and the environment. Environmental problems stem from market and policy failures. The international aspects of environmental problems reduce an individual country's ability to correct them. Harmonization which standardizes environmental requirements in different countries is not desirable, since it gives rise to inefficient solutions. Dealing with the consumption of imported goods which lead to environmental damage legitimizes product norms and standards, but to see at which level this is desirable – local, national or international – is not straightforward.

International environmental effects are also the result of production processes. Other countries may suffer damage either because the environment is abused in a manufacturing country (the equivalent of a manufacturing subsidy) or because of emissions which travel across geographical boundaries. The management of such issues requires international cooperation. This can be accomplished by common taxes and charge systems, and by coordination of international transfers and national measures in countries which lack the ability to pay. Trade barriers, on the other hand, are usually an ineffective instrument when trying to tackle environmental problems and risk being abused for other purposes as well as obstructing desirable cooperative efforts.

To allow spontaneous market solutions in the area of the environment and to lessen the danger of opening the door for capricious trade policy-making, it is essential that international environmental agreements and the multilateral trade system do not conflict. Since the costs of implementing environmental measures rise if they are introduced without careful preparation, it is important to be prepared – given that environmental concerns will eventually demand action. Unambiguous international agreements in the area of the environment would lead to a general spurt in efforts not to be left behind. A move in this direction would be facilitated by easily accessible and reliable information on how individual players affect the environment.

It is useful to look more closely at what an individual country can do, in terms of international cooperation and on its own, to address international environmental issues. Continued progress in terms of welfare today requires an upgrading of labour force competence, efficient handling of modern technology, flexibility and structural change to take advantage of new opportunities. International demand is weak for undifferentiated products with a low refinement value, while competition is tough, particularly from newly industrialized countries in East Asia, and potentially from other parts of the developing world and Eastern Europe. In this situation, a one-sided focus on cutting costs results in a gloomy perspective for the economy.

It is unavoidable that a growing environmental consciousness is creating significant adaptation problems for many industries. This is especially true for energy and raw material-based industries. Increased recycling of paper products seems, for example, to be affecting demand for wood within the near future in a negative way. On the other hand, the call for reduced carbon dioxide emissions and the need for alternative fuel for transportation may lead to higher demand for forests and forestry products from a slightly longer term viewpoint. Meanwhile, the development of environmental technology brings new possibilities in product as well as process development. Governments cannot and should not try to foresee which specific ventures will be likely to succeed, but can still act as a promoter and qualified purchaser in areas where environmental technology has a future. Thereby governments can contribute to economic expansion and new competitive advantages as well as a more favourable management of the environment.

It will most likely take time before effective international programmes are developed which compensate countries for the maintenance of life-supporting ecosystems of global importance. In connection with offering aid, however, there are ways to stimulate and facilitate indirectly a sensible resource use in poorer countries. Aid agencies need to increase their knowledge and administrative competence in this area. Efforts should be made to achieve a better synchronization of aid with the efforts made by private industry to spread 'environmentally friendly' technology to developing countries.

In the former state-planned economies of Eastern Europe, particularly in the former Soviet Union, there are a number of environmental time-bombs which can result in far-reaching consequences not only for the immediate area but also for the population of neighbouring countries. Until now, the West has been, for the most part, satisfied with determining what the problems are. This is an area where individual countries should more actively seek ways to take a stand with selective measures which can inspire more comprehensive ventures in cooperation with others.

Chapter 3

Trade Regulations – the Institutional Framework and Current Policy

There is a widespread misconception that free trade is desirable. It is not unusual that the GATT (General Agreement on Tariffs and Trade) is referred to as a 'free trade agreement' and the EU is referred to as a 'free trade zone'. These expressions imply that there would be no restrictions whatsoever on trade; trade would not be regulated at all.

Let us establish that there are no free markets or free trade in the true sense of the word. Markets (that is places where things are bought and sold) and trade (that is buying and selling) always function within different institutional frameworks. Legislation (competition legislation, taxation, and so on), trade regulations (GATT, bilateral trade agreements) and cultural aspects are examples of such frameworks. Free trade presupposes totally free (unregulated) markets and would mean a total lack of institutional frameworks. Consequently, there would be no taxes, no laws and no international agreements.

If free trade existed, it would be permissible to trade anything that someone else was willing to buy or sell, including people, narcotics, atom bombs and toxic waste, to mention a few of the more striking examples. From this perspective, it seems quite obvious that free trade is not something to be aimed for. This observation is also confirmed by the fact that countries all over the world have chosen to limit different forms of trade, for example within the areas referred to above.

What is generally meant by the phrase 'free trade' is *increased liberalization* of international trade, the removal of trade barriers which impede the flow of goods between countries, hindering an efficient utilization of resources and thereby impairing the general welfare. Although a freer flow of goods across borders is justifiable in some instances, there are obvious reasons why it is necessary to restrict trade in order to increase welfare. This is expressly recognized in GATT rules and EU regulations, since certain paragraphs grant member countries the right to introduce rules (for instance environmental regulations) which restrict trade in a number of ways.

This means that there is no truly free trade and no areas exist that are not covered by some trade restrictions. However, there are geographical areas within which countries have agreed that in principle it is permissible to sell a particular item everywhere, given that the item has been approved for sale in a country included in that area. The EU has, since 1 January 1993, represented such an area, as a further development of the original customs' union. Individual membership countries, however, again have the ability to adopt measures which restrict trade, if they have justifiable reasons, for example protection of the environment (the *Cassis de Dijon* Principle, see page 99).

The question is therefore not *whether* trade should be regulated, but *how* it should be regulated. This leads to the difficult question of what kind of trade should be restricted. How should an institutional framework be designed with regard to trade, in order to maximize welfare, both materially and otherwise? Which types of trade are desirable and which are not?

It is impossible to give simple and complete answers to such questions. The answers lie in what political systems different countries are prepared to accept, which means that the answers will differ considerably between countries. This result is quite natural considering the diversity of cultures and religions which exist around the world. There is nothing odd about the fact that there are different opinions as to what is good or bad, desirable or undesirable; on the contrary, it is very human.

There are, however, certain values and qualities which it is in the interest of everyone to preserve, namely the basic prerequisites for human life. Some examples are protection against ultra-violet radiation in the form of a functioning ozone layer, a normal climate and biological diversity which ensures genetic selection and prevents resilience (the ability to absorb a disturbance) from diminishing. In general terms, development which preserves such values is usually called sustainable development or ecologically sustainable development.

This chapter examines the opportunities available for the adoption of environmental protection measures which affect international trade within the framework of the GATT/WTO (World Trade Organization). The EU trade regulations are also discussed from the same perspective, albeit somewhat briefly. An overview of certain key functions in the NAFTA (North American Free Trade Agreement) is then presented. At the end of the chapter, we discuss how international trade regulations would need to be changed in order better to represent the demands set by the international community's goal of ecologically sustainable development.

GATT/WTO: TODAY'S PLAYING FIELD FOR WORLD TRADE

GATT was established in 1947, to try to achieve better organization of international trade, which after World War 2 was highly regulated, as it had been for specific situations during the war years.

Until the recent completion of the Uruguay Round and the establishment of the WTO, GATT had not been an independent organization, but was merely an agreement the intentions of which can only be interpreted by the member countries. The agreement has never formally come into effect, but the fact that GATT has existed and been applied since 1947 has resulted in GATT rules being regarded as binding.

Since its beginning, GATT has provided a forum for the simplification of trade regulations among the member countries. In 1992, GATT had 108 members, accounting for approximately 90 per cent of world trade.

In principle, the GATT agreement only regulates trade of goods and products, and not currently production processes.

GATT's objective is to promote economic development by removing various barriers to international trade.[116] The agreement gives detailed instructions for GATT countries' conduct in the trade policy arena, thereby regulating the conditions for how countries should act when designing trade policies. Although GATT does give detailed instructions, there remains considerable scope for interpretation.

Within the GATT framework, six general tariff negotiations were completed between 1947 to 1967. In principle, these aimed first to map out quantitative restrictions and customs duties, and then to minimize the levels of duties. This process continued from 1973 to 1979 during a further round of negotiations, the Tokyo Round. New agreements for levels and scope of customs duties, as well as new multilateral agreements regarding non-tariff measures, were finalized during these negotiations. The TBT Agreement (Technical Barriers to Trade) can be found among these agreements. Explicit use of the word 'environment' in a GATT context was first made in the TBT Agreement.

The Uruguay Round and the environment

The Uruguay Round began within GATT in 1986. It concerned custom duty levels and technical trade barriers as well as trade in services and intellectual property rights (such as patent rights), which are completely new areas of interest for GATT. The completion of the Uruguay Round in April 1994 also meant that a separate agreement on sanitary and phytosanitary measures was established, the SPS Agreement (Sanitary and Phytosanitary measures). This agreement specifically mentions the environmental issue, in approximately the same manner as the TBT Agreement.

116 Swedish Ministry of Foreign Affairs, Trade Department (1984) GATT, General Agreements on Tariffs and Trade

Aside from the TBT and SPS agreements, environmental issues were only indirectly addressed in the Uruguay negotiations. A technical working group (the EMIT-group; Environmental Measures and International Trade) was established within GATT in 1991. The EMIT-group had a mandate, a chairmanship and an organizational position which turned it into more of a governmental discussion-club than a forum for negotiations. Therefore, the trade-environment contribution from the EMIT-group was more educational than operative in respect of changing rules or interpretations of rules.

In the preamble to the agreement establishing the WTO, the contracting parties expressly recognized the objective of sustainable development, and the need both to protect and preserve the environment. In addition, GATT member-countries also decided to establish a sub-committee on trade and the environment. The sub-committee has an extremely liberal mandate (the decision to establish a sub-committee on trade and the environment, including the committee's mandate, appears in full in Appendix 5), which makes it possible to discuss almost any area of trade and the environment which the contracting parties feel would be appropriate. But perhaps most important is the fact that the mandate makes it possible to make recommendations on any modifications of the provisions to the multilateral trading system that may be required for promoting sustainable development. In practice, this means that the completion of the Uruguay Round, much criticized by environmentalists for not taking environmental aspects into account, actually provided a forum for negotiations on trade and the environment. With the establishment of the sub-committee on trade and the environment, a vehicle has been created for a possible transformation of the glorious words from the WTO preamble, 'to protect and preserve the environment', into action through re-negotiation of rules or interpretations of rules within the GATT system.

National regulations create comparative advantages

One of the central intentions of GATT is that any advantage, favour, privilege or immunity granted by any contracting party to any product originating in or destined for any other country applies equally to like products originating in or destined for the territories of all contracting parties. No country should discriminate against any other (Article 1, GATT, regarding the principle of MFN – 'most favoured nation'). In line with this principle, countries should not introduce trade restrictions which discriminate against foreign producers in relation to domestic producers (Article 3, GATT, the principle of 'national treatment'). The foundation for these two central GATT principles can be found in economic theory (see Chapter 2). The theory shows that overall material welfare improves, given certain assumptions, if countries trade with each other, and specialize their production in items they produce at the highest quality and best price compared to other countries. The fact that these conclusions are based on assumptions which need not be fulfilled in reality is discussed in Chapter 2.

Yet another dimension of GATT appears when discussing the content and interpretation of the phrase 'comparative advantages', namely national legislation and regulations. These are regarded by GATT as part of a country's relative competitive strength. GATT members have agreed not to introduce discriminatory customs duties for different goods. This makes it difficult to 'even out' differences in production costs by collecting duty on imports from countries which have, for example, lower taxes, fewer long-term requirements for labour protection or less-developed social welfare systems (these items are often referred to as 'self-induced costs').

It can, however, be strongly disputed whether, for example, the rapid destruction of a country's rain forests can be regarded as a comparative advantage for that country. The country can, in the short term, gain some export income as a result of low production costs, but in the long term, the conditions for a sustainable income source from the forest are undermined. It also impoverishes international resources such as the biological diversity which exists only in rain forests. It is believed in addition that the rain forest also represents an essential link in the sensitive interplay between the different factors which determine the world's climate. For reasons identified in Chapter 2, these aspects are not considered by the market.

For quite some time there have been discussions within GATT about the introduction of some form of social clause.[117] A clause of this nature would make it possible to discriminate against imports from countries which have what could be identified as insufficient labour protection conditions and therefore incur lower production costs. A social clause has so far not been introduced, but there is constant discussion of the subject.

There is a certain resemblance here with the case when certain countries allow inferior environmental protection with the aim of holding production costs down, thereby seemingly increasing their own competitive strength. Policy-making – or the lack of it – which results in the degradation of national and global environmental qualities is hardly a reflection of comparative advantage. Such policies may, in GATT terms, be referred to as hidden subsidies to industry, which, in principle, are forbidden according to GATT rules. As Franklin Roosevelt said in a 1937 speech to the US Congress: 'Goods which are produced under conditions which do not meet a rudimentary level of decency should be regarded as contraband and not be allowed to pollute the roads of international trade.'[118] On the other hand, it is obvious that countries cannot force their own priorities onto others.

Violation of GATT rules

In principle, GATT rules are binding, but there are a number of exceptions to which we will return later. It can be stated that no country is automatically 'guilty' if it breaks one of the GATT rules.

117 Molander, P (1992) 'Frihandeln ett hot mot miljöpolitiken – eller tvärtom?' Swedish Ministry of Finance, Expertgruppen för studier i offentlig ekonomi, Ds 1992:12
118 Worldwatch Institute (1993) *Annual Report*

Every country is obliged to notify GATT of all measures which may influence international trade. The notification procedure makes it possible for contracting parties to be sure that their rights, according to GATT, are not violated. In such cases, it is possible for countries to register complaints about the measures (whether they have merely been suggested or already put into practice) which they deem to infringe their rights. First, bilateral negotiations are to be initiated between the parties concerned. If no solution to the conflict can be arrived at by discussion then the plaintiff can request (from GATT) a panel be appointed to decide whether or not the GATT rules have been violated.

The panel consists of three experts with special competence in trade policy. The experts are appointed by GATT and need to be approved by the parties involved. The panel has the right to consult with each international organization deemed appropriate, to procure the specialized knowledge it requires. The accused country bears the onus of proof that it has not acted in a way which would constitute a violation of the GATT rules. The outcome will depend upon which paragraph of the GATT rules the plantiff is referring to.

Using the petition from the plantiff as a foundation, the panel must determine whether the GATT rules have been violated. The panel decision is not, however, equal to the ruling of a court of law, but it is sent to the GATT Council (one ambassador from each contracting party), which can unanimously adopt the panel's recommendation. There need be only one dissenting voice in the Council for the panel's decision not to be adopted. So far, only countries which are GATT members have had the mandate to interpret the rules of GATT. It should, however, be noted that the dispute settlement mechanisms of GATT have been significantly altered with the conclusion of the Uruguay Round and the establishment of the WTO. There is now a variety of different procedures for dispute settlement, depending on a variety of issues, for example within which area of competence under the WTO a panel is established.

Yet another complicating issue is the fact that pre-WTO GATT will continue to exist side-by-side with WTO, for some time to come. The reason for this is that all pre-WTO GATT member countries have not yet ratified the new WTO agreement, and it is far from certain when this double identity of GATT can be put to rest.

One practical aspect of this organizational schizophrenia is that the status of on-going panels within the pre-WTO GATT/WTO complex is highly uncertain. At the time of writing (Spring 1995) it is not obvious how or even if the question can be solved, since member countries seem to be in fundamental disagreement on the practical ways forward.

It should, finally, be said that it is not clear how a certain measure will be judged by GATT. There is a great deal of room for alteration to accepted practice regarding interpretation changes over time and different types of questions. One of many instances where varied interpretations can be made is the meaning of the words 'arbitrary' and 'unnecessary' in Article 20. An illuminating example, from Sweden, is the increasingly restrictive attitude to the use of cadmium. The current regulation was introduced in 1985 and is formulated as a general ban on the use of cadmium, with a list of the permitted exceptions.[119] Is this ban necessary? Is it arbitrary? The ban has never been tried in GATT, since no other member country has challenged the Swedish measure. However, this is not to say that Sweden could today decide similarly to ban the use of mercury without a dispute. It would depend entirely on whether or not a contracting party filed a complaint against the ban.

Generally, environmental measures have led to few formal deliberations within the GATT framework and even fewer panels.

GATT EXCEPTIONS FOR ENVIRONMENTAL PROTECTION

As mentioned earlier, GATT rules are, in principal, binding. There are, however, a number of exceptions which can cause the main regulations in GATT rules to be set aside. We will describe the only exception which at present can be invoked to apply trade restricting measures for environmental protection, namely Article 20.

Article 20 in GATT is often called the rule of exceptions. It stipulates that measures applied in a manner which would not 'constitute a means of arbitrary or unjustifiable discrimination between countries where the same conditions prevail or a disguised restriction on international trade' are permitted as follows (Article 20 appears in full in Appendix 3):

- necessary to protect human, animal or plant life or health;
- relating to the conservation of exhaustible natural resources if such measures are made effective in conjunction with restrictions on domestic production or consumption.[120]

It should be noted that the word 'environment' is not expressly found in Article 20. However, the text has (especially 20 (b)) come to be interpreted as general environmental protection, since protection of human, animal or plant life or health is in practice what we, directly or indirectly, wish to protect through environmental regulations.

As already observed, it is not obvious what the words 'arbitrary', 'unwarranted' or 'disguised restrictions' mean, and what the phrase 'countries where the same conditions prevail' implies in practice is not completely apparent either. We return to these phrases later.

Conditions for the approval of environmental protection measures

If a country adopts measures to protect the environment (necessary to protect human, animal or plant life or health) and the measures influence international trade, the measures should fulfil certain conditions in order to be compatible

119 Svensk författningssamling (SFS) 1987:5; Molander
120 Swedish Ministry of Foreign Affairs, Trade Department (1984)

with GATT rules.

- Is the environmental protection *legitimate*? The word 'legitimate' implies a requirement to prove that an environmental problem actually exists, that is, 'scientific justification'.
- Is the measure *necessary*? Are there other alternative measures which fulfil the goal of environmental protection, where trade is influenced to a lesser extent?
- Is the measure applicable in such a way that *foreign producers are not discriminated against* in an unjustified manner? Simply put, are domestic goods treated in the same way as foreign goods?
- Is the measure applicable in such a way that *no particular GATT member will be discriminated against* in relation to other GATT countries? Are imported goods, regardless of the country of origin, treated the same as native goods?
- According to current interpretation of GATT rules, the measure can only be introduced in order *to protect national territory*.

It is also important to state that GATT, in principle, only regulates the trade of goods and products. Production processes – how the goods are manufactured – are not covered. This means that it is not permitted, in general, to adopt measures against imports, based on how the goods are manufactured in the exporting country. If an importing country believes that a manufacturing country has not protected the environment sufficiently during the manufacturing process, the importing country does not have the right to introduce import restrictions against the goods by invoking environmental concerns, *according to GATT rules as they are formulated and interpreted today*. As a main rule, only product-related effects (that is, hazardous substances, insanitariness, and so on) can be invoked according to GATT.

Extraterritoriality: interfering in other countries' policies

Trade restricting measures which are directed against conditions outside individual countries, for example against production methods in other countries, are usually called extraterritorial measures. One probable reason why GATT rules only concede measures which are directed towards product-related effects and not towards production methods, is that GATT, in principle, could otherwise be in violation of current international law. According to international law, each nation state has sovereign right over its own territory. If GATT rules conceded measures against products which are based on judgements of how the product is manufactured in another country, it would result in a distinct interference in the sovereign right of nation states to make decisions. This would favour large, economically strong countries, which could then force smaller countries to change their policies. In discussions on trade and the environment, the phrase 'environmental imperialism' is sometimes used.

The fact that GATT rules only allow measures which aim at environmental protection within a country's own territory seems obsolete and limited, since environmental damage is increasingly transboundary. Perhaps the greatest threats to the environment that we know of today are those of climate change and depletion of the ozone layer. Both are examples where the effect is global, that is, areas which lie outside the legal jurisdiction of any particular country involved. To this already bleak picture can be added the rapid depletion of genetic diversity through the extinction of plant and animal species. Fish, birds and other wild animals do not recognize the borders we have created. The environmental gain from protecting a species in one country would, therefore, be reduced if neighbouring countries did not introduce similar protective measures. In this light, it seems quite obvious that the GATT rules need to be modernized in order to accommodate the needs of environmental policy-making, pursuant to the objective of sustainable development and aimed at protecting global environmental values.

It should be noted, however, that when interpreting Article 20 it is not evident that it can be applied only when taking measures to protect a nation's own territory. One trade analyst from the USA points out that an addition, which was made but not included in the agreement, *would have* restricted the use of Article 20 in this manner. At the beginning of the 1900s, there was also a tradition of making agreements with extraterritorial influence. During the decades before the GATT agreement was written, a great many agreements were aimed at protecting people and the environment, not just in certain countries but in all countries, for example regulations connected to animal trapping.

In general, import bans were introduced within the framework of international cooperation, in order to help export countries implement protective legislation. On a few occasions, countries acted either unilaterally or collectively, hoping to influence countries which were not part of the agreements. One example of the latter is the ban against the production and import of matches manufactured using white phosphorus. White phosphorus caused serious health problems during the manufacturing process, but not when the matches were used.[121]

Finally, it should be kept in mind that GATT is by no means a static instrument, but rather an institutional framework for international trade which constantly evolves over time. This framework, in itself, is an expression of the preferences that the contracting parties attach to differing political objectives. Since GATT is in constant development and change, it is possible to change the rules or their interpretation in accordance with changed preferences and political objectives in the member countries.

According to Article 21 (on exceptions regarding national security) it is permissible to introduce any trade measures if they are taken to protect interests regarding national security, such as the following:

121 Charnovitz, S (1991) 'Exploring the Environmental Exceptions in GATT Article XX' *Journal of World Trade* 25:37–56

- i) relating to fissionable materials or the materials from which they are derived;
- ii) relating to the traffic in arms, ammunition and implements of war and to such traffic in other goods and materials as is carried on directly or indirectly for the purpose of supplying a military establishment (see Appendix 3).

It is quite remarkable that it is permissible to take measures against the importation of products which are manufactured by prisoners (see Article 20, Appendix 3), or even more for the reasons stated in Article 21 above, while at the same time it is not permissible to take measures against imports of products manufactured in a way which damages the earth's shield against ultraviolet rays, an essential prerequisite to human life as we know it today.

The explanation to these discrepancies most likely lies in the fact that the GATT rules were formulated quite some time ago. According to information from the Ministry of Trade in Great Britain, the first draft of Article 20 was written in the early 1920s. At that time, trade of goods manufactured by prisoners was a great problem from the point of view of competition, while the substances which destroy the ozone layer had not even been invented. Article 21 was formulated in the aftermath of World War 2, when the Cold War had just begun and nuclear arms were becoming an issue of extreme international friction. Against this background, it is easy to understand why the current rules have been drawn up as they have. But in light of the pressing environmental challenges that the global human community faces today, it is also obvious that current rules do not properly match the environmental needs that have arisen during the last two decades. Environmental degradation recognizes no artificial borders. Therefore, a solution on how best to accommodate the environmental need of protecting both national and global environmental values must be actively sought within GATT.

Special agreement within GATT on product regulations and related areas

In connection with the conclusion of the Tokyo Round in 1979, the GATT rules were supplemented by an independent agreement from GATT, namely the TBT Agreement. This regulates the utilization of national technical regulations and standards which can influence international trade. TBT covers all types of products, including agricultural products, but not the trade of services. Today's decisions in TBT are concerned with product features. After the completion of the Uruguay Round, however, the TBT Code will also regulate characteristics which relate to production methods and include other issues, for example requirements for packaging and labelling.

The proposal for an agreement on sanitary and phytosanitary measures, the SPS Agreement, covers all measures which are taken in order to protect human, animal or plant life or health.

The point of departure for both agreements is that contracting parties should not be prevented from taking measures which are necessary to protect human, animal or plant life or health. Since such values directly relate to what we usually call environmental protection, TBT/SPS can be interpreted as including environmental protection as a legitimate use for technical regulations covered by the agreement.

However, this does not give *carte blanche* to the introduction of all measures developed for protecting the environment. In a manner corresponding to GATT, it is required that measures fulfil certain conditions before they are permitted according to TBT and/or SPS agreements:

- Both the SPS and TBT agreements contain comprehensive decisions concerning *transparency*, that is, information for interested parties (other contracting parties). Information should be available both prior to a measure being taken, and afterwards on how the decision was reached.[122]
- According to both agreements, measures taken to protect the environment or health should be *legitimate*, meaning that there must be scientific justification indicating that the measures will actually protect the environment or health.[123] This is a further example of the scope for interpretation which exists in GATT, since it is quite clear that scientists are not all unanimous on all issues.
- Both agreements are based on Article 20. However, there are stricter requirements regarding *non-discrimination* in TBT than in SPS.
- Both agreements contain requirements that the measure which is taken should restrict trade to a minimal extent, while the environmental goal is reached. In SPS, there is a supplement stating that the above applies, but what is technically or economically feasible must also be considered.[124]

As we have already mentioned, there are now regulations which cover production methods in both the agreements.[125] TBT and SPS consequently contain regulations on measures which should be enacted in other countries. The regulations are, however, restricted to measures which directly relate to product characteristics essential to the importing country (TBT) or measures which are necessary to protect health within the importing country's territory (SPS). An example of the SPS Agreement is the management of timber in export countries regarding measures for killing pine nematodes, since the spread of these worms to importing countries could cause damage to their domestic forests.

A final item worth special consideration in the TBT Agreement is the regulation regarding international standards. TBT states that where relevant standards exist or where their preparation is close to completion, the contracting parties

122 TBT § 2.9, 2.10 plus 10.1-10.7; SPS § 27 and Annex B.
123 SPS § 6; TBT § 2.2.
124 TBT § 2.2 and 2.3; SPS § 19 and 21.
125 TBT Annex 1, § 1; SPS Annex 1, § 1.

should use these or relevant sections of them, except in the case that such international standards or relevant sections of them may not be appropriate for the contracting parties. Examples include the protection of human health or safety, animal or plant life or health, or the environment; fundamental climatic or geographical factors; fundamental technological problems.

In practice, this will mean that, even if international standards exist, for example relating to permissible traces of insecticides in food products, each country is still free to introduce more stringent regulations. The TBT requirements regarding standards should thus be regarded as a floor rather than a ceiling.

GATT PANELS ON ENVIRONMENTAL ISSUES

There have been few panels regarding methods for environmental protection during GATT's nearly fifty-year history. Of the hundreds of panels which have been appointed, only a small number have addressed Article 20. Perhaps the most interesting and comprehensive of the panels concerned with environmental protection has been the USA's ban on importing tuna from Mexico.

The tuna–dolphin case

Schools of yellowfin tuna in the Pacific Ocean often swim beneath schools of dolphins, which result in the dolphins being caught and killed in nets meant for tuna. The USA assessed a law in 1972, the Marine Mammal Protection Act, which aimed to protect marine mammals such as certain species of dolphin. The law sets a ceiling limit on dolphin catches as well as the number of dolphins in relation to the number of tuna caught. This legislation now applies in USA territorial water and the economic zone; ships travelling in international waters and registered in the USA are also bound by this legislation. The law decrees an import ban on tuna fish which are caught using methods which do not comply with USA standards. The legislation offers an opportunity to ban the importation of all fish from a country which does not comply with agreements made in this area.

In August 1990 an embargo was placed on the import of tuna from countries not fulfilling US standards. The embargo first covered five countries, but finally was only concerned with Mexico since the other countries promised to fulfil the American demands. Imports of yellowfin tuna from Costa Rica, France, Japan and Italy, among other countries, were later prohibited.

Mexico protested against the measures and requested consultations with the USA in November 1991. When nothing came of these discussions, they requested a GATT dispute settlement panel in February 1991.

Mexico argued as follows:

- The American measures are in opposition to GATT's Article 11 (ban against quantitative restrictions) and Article 13 (ban against discriminatory measures connected to specific geographical areas).

- The possibilities to expand the embargo on all fish from countries which do not fulfil specific regulations regarding tuna are in opposition to Article 11.
- The marking of approved products (dolphin-safe) is discriminatory and in opposition to Articles 9 and 1.

The USA cited Article 3 in its answer and considered the treatment of imported goods to be no less favourable than the treatment of similar domestic products (see page 96 and Appendix 3). The USA also stated that even if the measures could be considered to be in opposition to Article 3, then they could be defended by citing Article 20 (b) or (g).

Mexico replied in its comment to the American statement that Article 20 refers to one individual country and cannot be used extraterritorially, that the affected dolphin species were not threatened with extinction according to the Washington Convention's Appendix 1, that the same species were influenced by other fish in the Northern Pacific Ocean which were not under the same restrictions and that it is generally doubtful whether the phrase 'exhaustible' can be applied to a population of living creatures.

The panel decision

In the panel report, published in August 1991, the American measures were found to be in opposition to Articles 3 and 11. The panel stated that only measures connected to the product are in agreement with the GATT charter and that the American measures related directly to *production methods*.

The panel also noted that GATT did not specifically say anything about the possibility, with the support of Article 20, of taking measures for the protection of the environment in areas outside a country's own territory (extraterratorial). The preliminary work and the aim of the 'exception clause' were therefore analysed, as were the potential consequences of the panel's interpretation for GATT. According to an early version of Article 20, measures could be taken to protect plants and animals if corresponding domestic protective measures existed in the importing country. This was considered by the panel to mean that the aim is to allow protection of the environment only within a country's own territory.

The panel further claimed that the measures to be taken by the United States could not be regarded as fulfilling the requirements of necessity. One of the reasons was that the technical design of the US measures was such that Mexican fishermen were faced with harder restrictions than US fishermen. The possibility of instituting environmental marking ('dolphin-safe') was not regarded to be in opposition to GATT.

The panel also emphasized that GATT rules offered substantial opportunities for member countries to institute measures which would restrict trade in order to protect the environment within their own territory. The panel also stated that if import restrictions based on differences in countries' environmental requirements were to be allowed then it would be necessary to define the limits for how and where this could occur so that abuse in the form of, for example, disguised restrictions could be prevented. If such possibilities are to be created, it would be better to change the agreement and simultaneously safeguard possible changes with limitations and criteria rather than to extend the present interpretation of Article 20.[126]

What happened afterwards?

In reality, the USA did not follow the GATT panel's recommendation, and the panel was never formally adopted. Mexico decided not to pursue the question further. In the spring of 1992, the USA and Mexico began negotiations for an agreement on a five year moratorium on the use of the much-discussed purse seine nets.[127]

Tuna–dolphin: panel 2

The European Union demanded a new panel against the measures taken by the USA. The measures the EU complained about primarily concerned the legitimacy of certain American secondary import restrictions on tuna and tuna-products from countries which had imported tuna or tuna-products from countries which were targeted by direct American import restrictions.

Without telling the whole story of the panel's report, the panel's interpretation of Article 20 may be a milestone in discussions on trade and the environment, and therefore deserves to be examined in detail.

The USA argued that if the panel would find both the primary and secondary import restrictions to be in violation with Articles 3 and 11, the measures taken would still be legitimate in accordance with Article 20(g) in the pursuance of preserving exhaustible natural resources. The USA argued that there are no requirements in Article 20(g) stipulating that the natural resources must be found within the national jurisdiction of the country taking the measures. Further, the USA claimed that the measures were taken in relation to limitations on domestic consumption and production, and that all other conditions mentioned in the beginning of Article 20 were fulfilled.

The EU, on the other hand, claimed that Article 20(g) is only applicable in cases when the natural resources, which are subject to preservation, can be found within the national jurisdiction of the country which is taking measures. The EU also argued that the American measures were aimed neither at preserving exhaustible natural resources, nor were applied to domestic production or consumption.

The panel first established that the motives of USA-regulations for preserving dolphins were such that they were in accordance with the general aim of Article 20(g).

126 GATT (1991) 'Panel Report' DS 21/R
127 French, HF (1993) 'Costly Tradeoffs. Reconciling trade and the environment'. In Worldwatch Institute Paper 113, p. 45

Second, after a comparison with Article 20(e), which concerns goods produced by prisoners, that GATT does not categorically exclude measures taken to influence activities which lie outside the national jurisdiction of the country taking the measures. The panel noted that the measures taken by the USA were aimed at changing the national policy of other countries, and that such changes were necessary for the import restrictions to have an effect on the preservation of dolphins.

The panel concluded that it was an open question whether or not Article 20(g) permits measures aimed at making other countries change their national policies. In the end, however, the panel argued that there was precedence within GATT to interpret the exemptions in Article 20 narrowly. Therefore, the panel decided that the USA import restrictions could not be legitimized with reference to Article 20(g).

Two important conclusions can be drawn from the panel's decision:

- In the dispute between Mexico and USA (see above) on the tuna–dolphin issue, Mexico claimed that a population of living animals could never be referred to as an exhaustible natural resource. The panel on the dispute between the EU and the USA clearly established that living animals, according to GATT, can be regarded as an exhaustible natural resource.
- In Paragraph 5.20 of the panel-report, it is clearly stated that it is not reasonable that Article 20(g) could only be referred to when taking measures to preserve exhaustible natural resources within the proper jurisdiction of the country which takes the measures. This opens up for discussion the question of under what circumstances Article 20(g) could be referred to when using trade-measures with an extraterritorial effect for the preservation of exhaustible natural resources. This issue has a direct link with the discussion on GATT and international environmental agreements, and how to find a way to accommodate trade measures taken in accordance with such agreements under GATT.

GATT AND INTERNATIONAL ENVIRONMENTAL AGREEMENTS

Since GATT was created in 1947, environmental problems have changed in character. The changes described in the previous section, that is from local to transboundary, and even global problems, have introduced a whole new dimension to environmental work. New solutions are required which demand distinctly intensified international cooperation, since no single country can solve the problems alone by taking measures in its own territory.

With these factors forming the background, negotiations on international environmental agreements have become an increasingly important part of environmental policy in most countries. Until now, the work has resulted in approximately 170 international environmental agreements being signed,[128] where countries undertake different long-term environmental measures depending on the level of development and other factors. An important element of these agreements is that industrialized countries often allocate money to international funds, which are intended to be transferred to developing countries, enabling them to comply with the international environmental agreements.

Barely 20 of the international environmental agreements contain measures which are directed towards trade (see Appendix 1). Decisions on the regulation of international trade have, in general, been introduced so that the effects of other measures will not be negated. An example of this kind of convention is the Montreal Protocol which is an agreement among 111 countries aimed at reducing the production and consumption of ozone-depleting substances.

In pace with the increase in the number of international environmental conventions with regulations regarding trade, the risk of conflicts between GATT rules and the international environmental conventions has also increased. As we shall see below, the legal relationship between international environmental conventions and GATT is not completely clear.

GATT and the Montreal Protocol

In the Montreal Protocol, trade in ozone layer threatening substances is regulated between countries which have signed the protocol as well as countries which have not. According to Article 4, member countries may not trade in those substances regulated in the Protocol (including products containing those substances) with countries which have not signed the Protocol. However, trade in regulated substances as well as products among those countries which have signed is permitted.

The motive for introducing this trade regulation was the risk that industries which manufacture substances that threaten the ozone layer could move their production to countries which had not signed the Montreal Protocol. Such a move might cause the member countries' decreased emissions to be nullified, since the environmental problem would simply be moved from one location to another. It does not matter where CFC is released; its influence on the ozone layer is equally negative.

Using GATT terminology, the Montreal Protocol's trade regulations can be described as a choice by the members to *discriminate* against non-member countries concerning trade in substances and products containing those substances, as regulated in the Protocol.

It is in this instance that problems can arise regarding the relationship between regulations in the Montreal Protocol and regulations in GATT. As previously described, GATT member countries have agreed not to treat another GATT country less favourably than the 'most favoured nation' (Article 1). One

128 Brown, L *et al.* (1993) *State of the World 1993*, Worldwatch Institute

potential problem between GATT and the Montreal Protocol lies in the fact that some countries signed GATT but opted to remain outside the Montreal Protocol (see Appendix 1). The following example illustrates the problem: according to GATT, Sweden has committed itself to treat South Korea as favourably as, for example, the UK, at the same time that Sweden has agreed, according to the Montreal Protocol, to discriminate against South Korea in relation to the UK when it comes to trade in substances (and products containing these substances) that are regulated by the Montreal Protocol.

A further aspect of the relationship between trade measures is that the present interpretation of GATT only provides for measures which are 'necessary to protect human beings, animal or plant life of health' (Article 20(b)) *within a country's own territory*. In the respective discussions within the OECD and GATT's Group on Environmental Measures and International Trade, the questioned has been raised as to whether measures to protect the ozone layer are really in agreement with the current interpretation of Article 20, that is, whether the ozone layer is actually a part of national territory.

At the same time, a conflict will not arise unless one GATT member brings an official complaint against another GATT member. Furthermore, no country has as yet questioned the Montreal Protocol.

It is not clear either how a potential GATT panel would rule on these questions. According to international law, a new agreement takes precedence over a previous agreement and a specific agreement takes precedence over a general one.

It is difficult to date GATT since it is constantly being re-negotiated. The fact that GATT has never formally come into effect, even though it has been applied for almost 50 years, further complicates the issue. The Montreal Protocol is, however, without doubt more specific than GATT. The Montreal Protocol also has more members than GATT. This will not however resolve a potential dispute, since a GATT panel can only consider whether or not the GATT rights of the contracting party bringing the complaint have been inadmissibly violated.

However, a large number of GATT member countries have signed the Montreal Protocol, and it is hardly in their interest to set a trap for themselves. GATT is not an independent organization but a system of rules which can only be interpreted by member countries. It is therefore unlikely that a measure taken in accordance with trade decisions in the Montreal Protocol would not be approved of by the GATT Council, even if a country which had not signed the Protocol brought a formal complaint.

The other international environmental agreements, where compatibility with GATT has been discussed, are the Basel Convention on the Control of Transboundary Movement of Hazardous Wastes and their Disposal, and the CITES (Convention on International Trade in Endangered Species of Wild Fauna and Flora). CITES came into effect in the early 1970s. There have been no instances of a GATT member country complaining that their GATT rights have been violated by application of the trade decisions in this environmental agreement.

GATT and Agenda 21

In November 1992 it was decided within GATT that Agenda 21 should be submitted to the EMIT-group in GATT for the discussion of relevant questions. Certain items (Chapter 2A in Agenda 21) will be examined by GATT's committee for trade and development. Some of the relevant issues have already been under examination within the standing agenda of the EMIT-group. At the same time, it is obvious that all trade and environmental issues in Agenda 21 are not covered under these points.

As already mentioned, a specific sub-committee on trade and the environment was created in connection with the completion of the Uruguay round. In the preamble of the agreement establishing the WTO, it is stated that members' 'relations in the field of trade and economic endeavour should be conducted with a view to raising standards of living, ensuring full employment and a large and steadily growing volume of real income and effective demand, and expanding the production of and trade in goods and services, while allowing for the optimal use of the world's resources in accordance with the objective of sustainable development, seeking both to protect and preserve the environment and to enhance the means for doing so in a manner consistent with their respective needs and concerns at different levels of economic development.' This means that the core message in Agenda 21 has actually been incorporated into the agreement of the establishment of the WTO.

There is some justification in giving special attention to one of the principles of the Rio Declaration, namely Principle 2: 'States have, in accordance with the Charter of the United Nations and the principle of international law, the sovereign right to exploit their resources pursuant to their own environmental and development policies and the responsibility to ensure that activities within their jurisdiction or control do not cause damage to the environment of other states or of areas beyond national jurisdiction.'

The principle embraces one of the central issues in the trade and environment debate, namely production processes in other countries and the opportunities for a certain country to take measures against conditions outside their own territory. Let us assume that a certain country contains a large proportion of the world's rain forests, which are of global interest and value in terms of biological diversity and world climate. Assume further that the same country has a long-term unsustainable forestry policy. According to Principle 2 of the Rio Declaration, that country has the right to exploit its own resources in accordance with its own developmental policies, but at the same time the country is in opposition to the second part of Principle 2, which states that activities within one country may not damage another country's environment, or areas which lie outside their national borders.

> *Provided there was broad support for the idea, GATT's member countries could decide to allow the imposition of import restrictions on exports from countries whose environmental regulations were considered to be 'inadequate' . . . It would be necessary to define inadequate environmental regulations and to develop procedures and criteria that would minimize the chances of abuse.*
>
> Trade and Environment, in *International Trade, 1990–1991*, GATT 1992

How will countries react when their interests are damaged but the interests lie outside their territories? What measures can countries take to protect their own interests and how important is it that they can do this in relation to the importance of upholding the principle that every country has jurisdiction over its own natural resources and development? It is obviously of the utmost importance to find a way to combine important principles of international law on decision-making rights over a country's own territory with the need to protect environmental values which are of global interest. This is clearly an issue of highest priority for the sub-committee on trade and environment to tackle.

EU TRADE REGULATIONS AND THE ENVIRONMENT

The Single Market aims to create economic growth and stability and thereby increase welfare. Member states' production resources are to be used as rationally as possible. Equivalent and undistorted competitive conditions should enable trade and exchange of goods, services, labour and capital, thereby furthering the goals of the community.[129]

The Treaty of Rome requires harmonization of national rules in order to dismantle barriers to goods and factor movements within the union. Such free movement cannot, for example, be hindered because of national regulations which raise different demands for a product's quality. Consequently, products can be marketed in other countries than the one they are produced in without the additional costs of adjustments during or after production.[130]

The EU used to be a customs union with a common border. The now fully integrated Single Market is concerned with all trade, and trade barriers cannot be applied either on goods which are exported or imported between countries. Any variation in domestic regulation is also banned. Once a product has passed the common border, it has free movement within the community and is covered by the Treaty of Rome's regulations on free movement.

In principle, the Treaty of Rome's regulations do not cover the member countries' tax systems. Tax regulations may not, however, create barriers for

129 Cecchini, P, *The European Challenge 1992: the Benefits of a Single Market*, Wildwood House, Aldershot, 1988

130 Cecchini, ibid

the free movement of goods, services, capital and services. Therefore, the treaty contains regulations directed towards fiscal discrimination. It is forbidden to levy tax on other member states' products above the level of direct or indirect taxes on similar domestic products or to, using taxation, indirectly protect certain products. If similar products are taxed at various levels, then that difference must be for objective reasons.

It may be recalled that the EC Court of Justice, in a fundamental ruling,[131] explains that any measure which can 'directly or indirectly, actually or potentially' influence trade negatively is prohibited.[132] This fundamental principle has since then, through the establishement of the Single Market, been extended to include not only goods but also services and production factors.

An important principle: 'Cassis de Dijon'

To the extent that specific directives (common regulations) for particular goods do not exist, the general EU rule is that if a product is accepted in one member state then it must be sold freely in other member states. Perhaps the best-known example of this question has been handled in the EC Court of Justice concerning a French liqueur, Crème de Cassis de Dijon. When the West German Government tried, at the request of their own liqueur producers, to stop the import of similar French liqueurs, they referred to West German alcohol policy regulations: the French liqueur, which has a lower alcohol content than the West German liqueur, was declared to be hazardous to some sections of the population as it would be abused by certain groups of people. This argument was not approved by the EC Court of Justice and West Germany was forced to withdraw its sales ban.[133]

Exceptions for environmental protection

The equivalent to GATT's Article 20 is Article 36 of the Treaty of Rome, which establishes that measures acting as trade barriers can be introduced on the basis of particular considerations. These considerations are similar to those which are named in Article 20 in GATT and are, with regard to consideration for the environment, described as 'protection of the health and life of human beings, animals or plants'.

Measures which are based on these considerations do not need to treat foreign and domestic goods alike. They may not, however, be arbitrarily treated differently or in a protectionistic manner, in the same way as the GATT rules. The principle that a measure must be necessary is also contained in the Treaty of Rome.

131 12/74 'Dassonville'

132 EEA-Agreement, Swedish Ministry of Foreign Affairs, Trade Department (1992)

133 Molander, P (1992) 'Frihandeln ett hot mot miljöpolitiken – eller tvärtom?' Swedish Ministry of Finance, Expertgruppen för studier i offentlig ekonomi, Ds 1992:12

A requirement for environmental measures influencing trade, which is found in the Treaty of Rome but not in GATT, is that *the environmental measure must be proportionately formulated.* In reality this means that the 'environmental gain' should be at least as large as the potential 'trade loss'.

Two of the most important court cases in the EC Court of Justice which are concerned with the environment are described here.

Danish bottles in the EC Court of Justice

In 1981 Denmark decided that all beer and soda must be sold in reusable bottles. The decision was made in the context of an old and well-functioning glass bottle recycling system which was threatened by increased sales of non-returnable beer bottles and other non-returnable bottles. The EC Commission considered the law to add greater costs to suppliers of imported goods than to suppliers of Danish goods and that the law thereby created a trade barrier preventing free movement of goods within the community. The case was handled by the EC Court of Justice in 1986.

The Commission claimed that Danish law conflicted with the basic regulation that all goods accepted in one member state must be accepted in all member states. Even though foreign bottles were not directly prohibited, foreign suppliers had greater difficulty creating a system to handle the reusable bottles. The Commission regarded this as discrimination against foreign companies. They also stated that the law was not in proportion to the environmental gain and that sufficient environmental protection could be reached by other means, for example voluntary collection systems and recycling instead of re-use.

The court ruling in September 1988 allowed Denmark to keep its law on obligatory re-use. The court confirmed that environmental protection is a consideration within the EC which can allow an exemption from the general rule of free movement of goods. If there is no specific EC law in that particular area, situations may arise where distortions of trade are caused as a result of the acceptance of different regulations in different countries. The court decided that a law stating that beverage packaging must be returnable was necessary in order to reach a high level of returns and that the measure was therefore 'in suitable proportion' to the goal.

However, Denmark's requirement for approval of bottle types was not accepted. After 1984, foreign manufacturers were required to sell goods in bottles which were accepted by Denmark or in their own bottles, but for test periods only and in limited quantities. The motive behind approving bottles was that it is not possible to have an effectively functioning return system (with high levels of returns) if there is a wide variety of types of containers on the market. By controlling the types of bottles, the varieties could be restricted. The court found that a system of recycling non-acceptable bottles does not guarantee a maximum recycling level but does protect the environment. As a result of this ruling, very few beverages are imported into Denmark. The court thereby found that the environmental gain from an obligatory approval of bottle types was not in proportion to the substantial drawbacks for foreign suppliers.[134]

Trade in waste: the Walloon ruling

In 1985 the Belgian Walloon region introduced regulations prohibiting the storage and deposition of waste from other countries or regions of Belgium. The EC Commission requested Belgium to appear in the EC Court of Justice and stated that the country had not fulfilled its commitment according to the Treaty of Rome's Article 30 and 36 in addition to several different directives dealing with waste which exist within the EC.[135]

On 9 July 1992 the EC Court of Justice confirmed that:

- Waste should be regarded as goods. Recyclable waste has a commercial value, but even non-recyclable material which is transported over a border and is an object for business transactions should be considered as goods, regardless of how the transaction may appear. It is, however, difficult to distinguish recyclable waste from other types of waste.
- It is not permitted to introduce general import prohibitions of environmentally hazardous waste under the 'exception' Article 36, since there is harmonization legislation in this area.[136]
- The Walloon situation is exceptional, with an abnormally large influx of waste from other regions. The overwhelming need for environmental protection can therefore justify an exception from Article 30.
- For an exception to be granted according to Article 36, the measures must not discriminate. The court ruled that it should be possible to uphold particular regulations for waste from outside the immediate district, since waste disposal management is a special process. According to Article 130 r2, environmental damage should be rectified at source, and the court stated that waste should be dealt with as close as possible to its place of origin so that the need for transportation is limited.

The conclusion from the Walloon ruling is that it would not be in accordance with EC legislation to introduce a general import ban for environmentally hazardous waste, since there is already a harmonized EC law in this area. However, there are possibilities to stop individual shipments of toxic waste.[137]

EC waste policy after the Walloon ruling

Since the Walloon ruling, EC waste policies have been amended. The directive

134 Case 302/86. Commission of the European Communities *v* Kingdom of Denmark (1988)
135 General directive on waste 75/442, directive 84/631 on surveillance and control of transboundary transports of hazardous waste
136 Directive 84/631
137 Christensson, B (1992) The Walloon Ruling – EU ruling on waste, c-2/90

on transportation of waste[138] was replaced by a new directive in late 1992.[139] The new directive confirms the principle of subsidiarity. It states that the transportation of waste should be avoided when possible and each country should be responsible for its own waste. Each country should also be able to refuse to accept waste from other regions. The directive follows the Basel Convention, which states that the export of hazardous waste is only permitted when the government of the importing country has accepted it. The new directive is based on the Treaty of Rome's Articles 130s and 130t, which state that the harmonized environmental law specifies a minimal level but individual countries can introduce more stringent measures if desired. The earlier directive was founded on Article 100a, which states that legislation is completely harmonized and that individual countries may not introduce more stringent measures.[140]

In 1991, the general waste directive was modified.[141] The legal basis for this modification is Articles 130s and 130t. The EC Commission then suggested that Article 100a could instead be used as the legal basis. In March 1993, the EC Court of Justice confirmed that, since the environmental goal is the main aim of the law, then Article 130s, which is regarded as the embodiment of EC environmental policy, should apply. The aim of the general waste directive is to restrict the production of waste, to encourage recycling and returning, and to guarantee that member states manage their own waste in addition to restricting the transportation of waste. This emphasizes the principle of subsidiarity and the self-supporting principle: the community as a whole should take care of its own waste with each member state striving towards that goal. The EC ruling states that regarding the waste issue, the subsidiarity and self-supporting principles have greater validity than the interests of free trade.[142]

A NEW GENERAL SYSTEM OF PREFERENCES

During the autumn of 1994, a new General System of Preferences (GSP) was negotiated and decided upon within the EC. The old system with quotas was abolished, and replaced by tariffs.

A completely new dimension of the GSP system was that it entails two elements of conditionality, concerning certain aspects relating to labour and environmental conditions.

Firstly, from 1 January 1998, special incentives in the form of additional preferences (that is, lowered tariffs) can be made available to countries meeting eligibility requirements under the GSP. The condition for receiving such duties

138 Directive 84/631
139 Official proviso 93/259
140 Jansson, M (1993) EG-regler om handel och avfall, Promemoria, Swedish Ministry of Foreign Affairs, Trade Department (No. 3)
141 Directive 75/442 was replaced by directive 91/156
142 Bergman, P (1993) EG-domstolens avfallsdom, Promemoria, Swedish Ministry of the Environment and Natural Resources

is that the country in question submits a written request documenting that it has adopted and actually applies domestic legal provisions incorporating standards of the International Labour Organization Conventions No. 87 concerning Freedom of Association and Protection of the Right to Organize, No. 98 concerning the Application of the Principles of the Right to Organize and to Bargain Collectively and No. 138 concerning the Minimum Age for Admission to Employment.

Second, from 1 January 1998, special incentives in the form of additional preferences can be made available to eligible countries under the GSP which submit a written request documenting they have adopted and actually apply domestic legal provisions incorporating the content of International Tropical Timber Organisation standards concerning sustainable forest management.

The new GSP system contains both a social clause and an environmental clause. During the negotiations of the new system, however, several member countries expressed their unwillingness to incorporate, into a trade agreement, conditionality regarding other policy areas. Therefore it still remains to be seen how the system will work in reality. There are no guarantees for receiving lower duties, since the texts only stipulate that the duties can be lowered.

From the Treaty of Rome

> *Article 30: Quantitative restrictions on imports and all measures having equivalent effect shall . . . be prohibited between Member States.*
>
> *Article 36: The provisions of Articles 30 to 34 shall not preclude prohibitions or restrictions on imports, exports or goods in transit justified on grounds of . . . protection of the health and life of humans, animals or plants . . . Such prohibitions or restrictions shall not, however, constitute a means of arbitrary discrimination or a disguised restriction on trade between Member States.*
>
> *Article 130 r2: Action by the Community relating to the environment shall be based on the principles that preventive action should be taken, that environmental damage should as a priority be rectified at source, and that the polluter should pay. Environmental protection requirements shall be a component of the Community's other policies.*

The entirety of Articles 130 r, 130 s and 130 t can be found in Appendix 4.

ENVIRONMENTAL PROTECTION AND FREE TRADE AREAS

A development which has become more definite during the last few years is the growth of ever larger free trade areas, in terms of sheer geographical area, as well as the depth of the market integration. The two foremost examples are the establishment of a Single Market within the EU, and the realization of NAFTA between the USA, Mexico and Canada. There are far advanced movements in this direction among countries in the developing world.

The definition of the phrase 'free trade' is, as we have so exhaustively discussed, not totally clear-cut, since regulations within the EU's inner market and the NAFTA agreement offer opportunities for barriers against the free movement of goods due to, for example, environmental reasons. However, the aim in both cases is undoubtedly to encourage as free a flow of goods as possible between countries.

The genesis and development of free trade areas (to use this term from now on) can have both positive and negative effects on the environment, depending on how regulations are formulated. The effects are clearly positive if the trade agreement is formulated such that the strictest applicable protection regarding, for example, the ban of goods containing hazardous elements should be applied to all countries ('upwards harmonization'). The implementation of such a prohibition may be somewhat problematic, depending on which countries are included and their respective development level. It can be more difficult for poor countries to apply strict protection levels than their wealthier neighbours. A change-over to stricter protection levels could take place by building some form of assimilated funds where wealthier countries in a given region donate money which the poorer countries can use to attain higher levels of environmental protection.

On the other hand, if regulations within a free trade area were drawn up so that the lowest applicable level of protection was allowed for the entire area, it would have a negative effect on the environment. However, it is hardly likely that the regulations would be created in such a way, since a number of countries are, after all, striving for a higher level of environmental protection. An example of this manifested itself in the Rio Declaration and Agenda 21.

Naturally, there are a number of possibilities which lie somewhere between these two extremes; where some levels are raised and others are lowered. It is also possible to formulate regulations in trade to provide minimum standards of protection. An example is a minimum requirement on the maximum amount of pesticides allowed in food products, and countries are free to raise that standard if they so desire. It should be pointed out, however, that the more requirements there are, the less 'free' the flow of goods is likely to be.

Another environmental aspect of the liberalization of trade is whether it will lead to higher economic activity (higher total production) and a changed direction for production. Another consideration is whether sufficiently stringent environmental measures are being taken to combat environmental effects such as increased use of land, increased transportation, increased air and water emissions, and so on. According to the EU's own research into internal market influences on the environment, it can be estimated that transportation via highways and roads will increase by 30 to 50 per cent.[143]

Trade influences price relations in both export and import countries. Since prices influence the use of resources, different prices create new production and consumption patterns. This has indirect consequences for the environment. Observers have, for example, warned that NAFTA can lead to more corn being cultivated in the USA and less in Mexico, forcing many small Mexican farmers from the market. It is predicted that beef production will in part move southward, resulting in the increased destruction of forested areas by creating new pastureland.[144] These aspects were discussed in the introduction and will not be further explored in this chapter.

In connection with the establishment of NAFTA, Mexico, the USA and Canada introduced (individually and as a group) analysis of the environmental consequences of the free trade agreement. This was probably the first time an analysis of this type has been carried out. Without going into the details of the result, it can be said that the agreement contains a number of 'safeguards' with regard to environmental protection. The safeguards which are of principle interest here are:

- NAFTA interprets Article 20 of GATT as including environmental measures (that is, the word 'environment' is specifically listed, which it is not in the GATT agreement) necessary to protect the human, animal or plant life or health, as well as measures which aim to preserve living or non-living exhaustible natural resources (compare to GATT Article 20 as discussed above).
- The contracting parties have the right to decide themselves which environmental protection levels they will introduce or maintain.
- All of the partners will work for a unified upwards harmonization of environmentally related standards (upwards harmonization).
- If a dispute concerning environmental standards should arise, then the burden of proof lies with the party bringing the complaint (the reverse of the situation in GATT).
- Reports on resolution of disputes which concern environmental measures will be published 15 days (at the latest) after the report is sent to the NAFTA Commission.
- All the parties involved pledge to strengthen the observance of environmental laws and regulations.
- Trade decisions in CITES, the Basel Convention and the Montreal Protocol (see above) take precedence over NAFTA decisions.

It should therefore be absolutely clear that the NAFTA Agreement, *as such*, is positive for the environment, since none of the current standards will be low-

143 French, HF (1993) 'Costly Tradeoffs. Reconciling trade and the environment'. In Worldwatch Institute Paper 113, p. 37

144 Ritchie, M (1992) 'Free Trade versus Sustainable Agriculture. The Implications of NAFTA' *The Ecologist* 22:221–8

ered and countries will be working for a successive and shared tightening of requirements. It remains to be seen whether the undertakings will be observed.

GROUND RULES FOR ECOLOGICALLY SUSTAINABLE TRADE

In discussions on trade and the environment, it is often stated that environmental measures should be applied in order to solve environmental problems, and trade policy measures should be used to solve trade policy problems. The implicit assumption is that each policy area exists independently of the other, therefore measures should only affect their 'own' area. This line of reasoning is certainly supported by some proponents of trade policy who fear to mix up trade policy issues with environmental issues. Measures which influence trade should thus not be used to protect the environment.

Such an outlook, for reasons already described, results in an untenable position. Today's reality is characterized by increased internationalization and an integration of politics and economics. Trade is an integrated part of today's production and consumption structures. It is not feasible to view trade separately, as if it were not linked to environmental damage resulting from transportation, production and consumption. Trade lubricates the economy and transportation lubricates trade. Concisely put: without transportation there is no trade, in principle, and without trade, in principle, there is no economy (production and consumption).

What is therefore needed to solve the world's pervasive environmental problems is increased cooperation between, for example, trade policies and environmental policies, so environmental concerns are included in the ground rules developed for international trade. In Sweden, this 'cooperation' between different policy areas is normally called sector responsibility. In reality, this implies that all policy areas must consider the broader consequences of measures within their particular policy area; the Minister of Transport is responsible for environmental effects within the transportation area, the Minister of Agriculture is responsible for environmental effects in agriculture and so on, with the Minister of the Environment acting as the spider in the web.

The same outlook is apparent in the EU's fifth environmental management programme and Agenda 21. It is stated that environmental consideration must be built into all decision-making processes in society, since most decisions can potentially influence the environment.

For this actually to happen, the *direct* effect of trade regulations on the environment and environmental policy (measures introduced to promote environmental protection), and the *indirect* effect of trade regulations on the environment must be taken into account. (For example, NAFTA could lead to increased soil erosion in Mexico, and realization of the EU's inner market could increase road transportation by 50 per cent.) The environmental consequences of trade agreements must be analysed if the indirect effects of trade regulations are to be considered. This was rare before the mid-1990s, largely because there was no obvious need clarifying potential. Where negative effects can be identified, measures must be taken to avoid their occurence.

Establishing methods to analyse the environmental consequences of trade and its agreements is an essential task for our future.

In relation to the direct influence of trade regulations on the environment and environmental policy (that is, the opportunity to implement measures for environmental protection which influence trade), there are several ways to approach the problem. It is quite clear that measures taken by individual countries to fulfil their commitments according to international environmental conventions cannot be constructed as opposing GATT for reasons of legitimacy (see page 100). In other words: the GATT rules or interpretation of the GATT rules must be changed, so that it is completely clear that measures taken in accordance with international environmental conventions cannot be regarded as opposing GATT.

One of the central questions in the complexities surrounding environmental and trade policy is the possibility of protection outside the territory of individual countries, given that there is no such international environmental convention. Should individual countries have the right to introduce measures which influence trade in order to protect the environment in other countries? Should individual countries have the right to introduce measures which influence trade in order to protect the value of the environment which can be attributed to global commons, when environmental destruction is a result of production processes in other countries?

One of the most important tasks in the future will probably be to identify the conditions under which it will be legitimate to introduce trade barriers in reaction to environmental damage resulting from production processes. From an environmental viewpoint, it is strange to legitimatize regulations against products manufactured by prisoners, while being prevented from banning the trade of products manufactured in a manner which threatens the earth's protection against ultra-violet rays. Scope should be provided for such measures within a framework for international agreements, but it cannot be denied that in special cases, where environmental degradation is advancing, countries may unilaterally need to stop trade in goods stemming from environmentally damaging production processes.

However, it is important to identify the real need for trade measures to avoid environmental measures in certain countries being offset as a result of other countries' insufficient efforts to protect the environment. At the same time, it is absolutely essential that clear environmental regulations are established within the framework of trade regulations. For small economies, clear ground rules are especially important, in order to eliminate the risk of ending up with a situation where 'might makes right'. Furthermore, it is vital to have clear and well-functioning trade regulations to prevent the occurrence of conflicts and acts of reprisal, which in the long run can be a breeding ground for much more serious conflicts than 'mere trade wars'.

> *Overall, it would be reasonable to argue that the GATT is more in need of protection from poorly reasoned demands for reform based on environmental arguments, than the environment is from the rules of the international trading system.*
>
> Low, P (1992) 'International Trade and the Environment' World Bank Discussion Paper 159, p.12.

The situation must be avoided where individual countries adopt measures on their own as a result of a lack of opportunities for international cooperation (whether trade policy or environmental policy). Therefore it is of the utmost importance that well-defined policies are created with sufficient scope for environmental measures within central trade regulations. This must make it possible for countries to adopt measures which are necessary in order to achieve long-term, ecologically sustainable development. If this does not occur, it is only a question of time before environmental degradation has gone so far that environmental issues may develop a dimension of national security policy. Once environmental issues are considered matters of national security, Article 21 in GATT could be invoked for environmental protection measures. It is, however, in the interest of the entire human race to solve the problem before it has advanced to that point. This requires world governments, as well as the negotiators of future trade agreements, to act with a great deal of responsibility and tackle seriously the challenge that this scenario presents.

Chapter 4

The New Playing Field – Towards Sustainable Development

Global integration, the increasing exploitation of resources, and growing environmental problems have highlighted the dependence of the global community on functioning ecosystems and ecological services.[145] Economics and trade must therefore be viewed from a holistic perspective which clarifies the link between nature's life support systems and economic activity.

The importance of integrating environmental issues in decision-making processes was emphasized in the Rio Declaration and in the EU fifth programme for environmental action. 'Cooperation' between various policy areas is often referred to as sector responsibility. In practice, this means that, in the present world with its jointly determined ecological–economic systems, all policy areas need to take into account the far-reaching consequences of measures implemented in their own fields. This also applies to trade policy.

Trade is a self-evident element of today's economy, and a key driving force in the global integration process. Therefore trade cannot be treated as a separate issue.

> ***With such looming problems as global warming, deforestation, or biodiversity loss, at least the issues are clearly defined. Trade, on the other hand, cuts across all those problems . . .***
>
> French, H F 'Costly Tradeoffs. Reconciling trade and the environment' World Watch Institute Paper 113, p.5.

Our book has presented an integrated ecological–economic view of overlapping issues in trade and environmental policy. The analysis consists of three main areas: objective, means and scope for implementation.

The *objective* is ecologically sustainable economic development. Our welfare

145 Holling 1994 (see note 34); Folke, Holling, Perrings 1995 (see note 36)

and survival, and a functioning economy, depend on nature's life support systems and the ecological services that they produce. A necessary condition for the production and consumption of goods and services to be sustainable is that the process does not impede the ability of the ecosystem to recover from disturbances. Consequently, the ecological system provides the framework for human activity.

Because of the true uncertainty which exists with regard to ecological thresholds and limits, it must be possible to undertake measures to protect the environment even when the necessity is not completely evident; it is too late to prescribe medicine when a patient is already dead. Individual countries or ecosystems cannot be seen as isolated from each other. 'Local' environmental problems should, to an increasing extent, be seen and dealt with as transboundary and global problems. The challenge is to be far-sighted and proactive – instead of being forced to respond by a deteriorating situation.

Given the ecological framework, economic analysis can be used as a *means* of influencing the players in society and of formulating rules for trade which will lead to efficient use of all society's scarce resources, in response to increasing ecological scarcity. In practice, this involves different types of policies, aimed at ensuring that, as much as possible, production processes and products which damage the environment carry the true costs of their actions.

An integrated ecological and economic analysis sets the conditions for ecologically sustainable development, establishing a new playing field. The feasibility of *implementing* the specified policies depends on how the institutional framework, in this context primarily GATT and EU regulations, is formulated and how it develops in the future. Whether they will change in accordance with the framework provided by the new playing field depends largely on our ability to change the view that humanity is superior to and independent of ecosystems. Such a change will be made easier if processes that are difficult to grasp, such as depletion of the ozone layer, become visible so that people are confronted by the significance of these processes in their day-to-day decision making.

It is not useful to blame certain actors, find scapegoats, or attempt to stop all economic advance, adopting the view that the limits have already been reached. This will only delay the process of change and create further deadlocks. Rather, the key to a solution lies in opportunities to change the behaviour of society *vis-à-vis* the ecosystem. There is no point in trying to stop the train. Instead it should be switched to a new track, where ecology and the economy work together.

GROUND RULES FOR SUSTAINABLE DEVELOPMENT

Signalling 'the new ecological scarcity' to the actors on the market

According to economic theory, liberalization of international trade may lead to long-term sustainable development. A precondition for this is that environmental values are included and that the cost of damage to the environment is borne by the products and production processes, that is, that the 'new scarcity' is signalled to the economic actors in society. In practice, we are far from such a situation. On the contrary, few countries have begun the internalization of environmental costs. The environment is continuing to deteriorate rapidly. Most indications are that environmental costs are increasing rather than decreasing. It is thus difficult to provide a simple answer to the question of whether liberalization of international trade is harmful or beneficial to the environment.

Two principles from the Rio Declaration (12 and 16) deal explicitly with trade and the environment. They recommend that states should promote an open international economic system which can contribute to economic growth and sustainable development. An open international economic system and economic growth, however, will not, by themselves, result in sustainable development. It is when an open international system is formed in conjunction with its ecological resource base, and when economic growth is based on this interaction, that environmental damage can be avoided and management of the ecosystem can proceed in a constructive and sustainable manner. Trade and growth need proper institutional frameworks.

The Rio Declaration states further that internalization of environmental costs and use of economic instruments and policy are to proceed so that international trade and investment are not distorted. It is, however, unavoidable that internalization of environmental costs will in many cases result in significant changes in patterns of production and consumption. This will have repercussions on the present system of trade. Since trade is based on market prices, it will change if environmental values are internalized. The purpose of internalization is specifically to promote more efficient use of society's resources, and influencing markets to include increasingly all of society's costs and benefits.

To establish this new playing field, the indirect effects of trade regulations on the environment must be taken into account. Environmental impact assessments of trade agreements must therefore be upgraded and refined, so that the negative effects of new trade agreements can be predicted and avoided.

It is often claimed in the debate on trade and the environment that trade measures are not effective and that other measures should be implemented to address the actual causes of environmental damage. While this is correct in theory, in practice it would mean that environmental gains from measures aimed at national production and consumption would be reduced through imports from countries lacking sufficient environmental standards. It would also mean that countries could not influence domestic environmental damage, when the causes of the damage originate outside their borders. Trade barriers can be an effective means to exert various forms of pressure on countries. Their effect, however, depends on how measures are formulated in individual cases and the extent to which they are viewed as meaningful by the various nations, in other words how rules are used and observed in practice.

Developing and intensifying international cooperation

Introduction of identical environmental standards in all countries will not be efficient, since natural and economic conditions and human values vary from country to country. A situation where countries draw their own rules, however, may be quite undesirable. Many countries decide, consciously or otherwise, to avoid internalizing environmental costs, which corresponds to indirectly subsidizing environmentally damaging activities. For example, this may happen when environmental rules or charges are avoided for trade policy reasons. The problem of indirect subsidies is particularly relevant to transboundary environmental effects of industrial pollution. This is due to the fact that not only the exporting country but also many other countries suffer the environmental effects, so they too are involuntarily subsidizing environmentally damaging industrial production.

Transboundary environmental problems can only be solved by international cooperation. It is in the interests of all nations to solve environmental problems, but the specific interests of different countries are far from identical, since both the costs and benefits of tackling such problems differ between countries. Therefore the key to truly effective forms of cooperation is to take the interests of different countries as the starting point and to coordinate them so that all countries are willing to participate. This may mean that countries commit themselves to differing levels of involvement. Cooperation must be defined so that involvement at different levels, both nationally and internationally, prepares a mutual way forward and obliges separate nations to follow agreements. In order for developing countries to raise environmental standards, transfers of different types will also be necessary in many cases.

Act now!

The GATT/WTO rules are entirely based on international law, and do not permit an importing country to discriminate against goods in terms of how they are produced or, with the help of trade barriers, to attempt to influence the environmental policy of another country. Under international law, each nation has the sovereign right to decide on matters concerning its own territory and to form its own individual policy. When an environmental problem originates in another country, the only course of action currently available is to attempt to persuade the culprit nation to stop polluting the environment. International law, however, is not adapted to current conditions, where one country can, by its environmentally damaging behaviour, impair fundamental living conditions in another country.

Worsening environmental problems are making the present situation untenable. Fundamental precepts of international cooperation must therefore be re-examined. If environmental conditions continue to undergo rapid change, demands for the introduction of trade barriers will undoubtedly increase. The GATT/WTO rules, or their interpretation, must therefore be amended so that measures adopted in conformity with international environmental conventions are not seen to be in conflict with GATT/WTO. Negotiations should then start at once on the creation of rules which allow the introduction of trade measures against products from countries with unacceptably low standards of environmental protection. It is particularly important that measures can be taken against production processes which inflict global environmental damage.

These measures must be implemented through international cooperation. Otherwise, there is a risk that environmental arguments will be (ab)used, especially by large countries, to motivate protectionist measures in general. Circumvention of international conventions is mainly at the cost of small countries. But it is also in the interest of larger countries to contain disorder across the world market, otherwise mutual confidence between states, and opportunities for international cooperation will deteriorate and there will be a heightened risk of trade wars. Well-defined international ground rules for world trade are therefore required, which fully account for the dependency of world trade on life-supporting ecosystems. In other words, it is in the interests of both trade and environmental policies that international rules and regulations for trade are developed in a way which accelerates the process towards ecologically sustainable economic development.

CONCLUSIONS AND RECOMMENDATIONS

- A society that neglects the environment moves closer to ecological thresholds and boundaries thereby reducing the scope of socio-economic development.
- In pace with continuous population growth and an expanding world economy, natural capital is increasingly becoming a limiting factor – 'the new scarcity'. Local environmental degradation is increasingly becoming a transboundary issue.
- Institutional frameworks for economic development must be introduced which account for the genuine uncertainty of the limits and thresholds of global ecological systems.
- To the greatest possible extent, those responsible for inflicting environmental damage should bear the costs of the damage. Otherwise environmentally harmful activities are subsidized by society.
- Trade magnifies economic development, regardless of whether or not economic development is sustainable.
- Trade is not the primary cause of environmental problems. Therefore the source of the problems needs to be rectified. To the extent possible, market forces should be engaged for this purpose.
- Trade barriers aimed at counteracting the environmental effects of production within another country's territory may be justified to facilitate an effective domestic environmental policy. Such measures, however, should be based on multilateral agreements. Unilateral measures should be precluded.

- Individual states should have the right to introduce product norms and standards which regulate the local environmental effects of consumption. When the rules are harmonized to facilitate the free flow of goods within a free trade area, individual countries should not be forced to lower their environmental standards.
- International harmonization of environmental standards is largely inefficient, although minimum requirements are justifiable.
- Transboundary environmental problems must be solved by international cooperation and the coordination of international and national measures in different countries. It should be recognized that the environmental standards of different countries do not have to be identical and certain countries should receive financial support in order to comply with international agreements, but their own incentive to do so must be safeguarded.
- The GATT/WTO rules must be further developed in order to allow measures to be taken by individual states which affect international trade but which comply with international environmental agreements.
- Before trade agreements come into effect, environmental impact assessments should be carried out and corrective measures implemented in cases where serious environmental damage is identified.
- Internalization of environmental effects leads to new investments and changes the conditions for international competition. Governments have an important role to play in promoting the development and spread of environmentally-friendly technology.
- National and international measures are required to increase access to reliable information about how individual companies and institutions affect the environment.
- The lack or absence of democracy, freedom of speech and freedom of the press contribute to environmental damage in many countries.
- Appropriate institutional frameworks for trade can accelerate the shift to sustainable development. This is in the interests of effective trade policy as well as effective environmental policy.

Appendix 1

INTERNATIONAL ENVIRONMENTAL AGREEMENTS WITH TRADE PROVISIONS

The date next to the title of the agreement is the date on which it was signed.

Convention Relative to the Preservation of Fauna and Flora in their Natural State, 1933

The aim of this agreement is to preserve natural flora and fauna of the world, particularly in Africa, by means of national parks and reserves, and by regulation of hunting and species collection. The agreement includes a prohibition against the import and export of trophies, unless the exporter is given a certificate permitting export.

International Convention for the Protection of Birds, 1950

The objective is to protect populations of birds, particularly migratory birds, from extinction. Ten Western European states have signed the agreement. It includes a prohibition on the import, export, transport, offer for sale or sale of live or dead birds killed or captured during the protected season, or of eggs, their shells or their broods of young birds in the wild state during the breeding season.

International Plant Protection Agreement, 1951

The objective of the convention is to maintain and increase international cooperation in controlling pests and diseases afflicting plant products. The undersigned agree to regulate strictly the import and export of plants.

European Convention for the Protection of Animals during International Transport, 1968

The parties agree to fulfil the provisions of the convention governing the international transport of animals.

Convention on International Trade in Endangered Species of Wild Fauna and Flora (CITES), 1973

The trade of certain species is regulated in order to protect threatened animals from extinction. Certain species cannot be traded while the trade of other species is authorized by export and import permits. Endangered species are listed in three classes. Species threatened with extinction, appendix 1, species that may become endangered unless trade is strictly regulated, appendix 2, species that a party identifies as being subject to regulation within its own jurisdiction and as requiring international cooperation to control trade, appendix 3. The agreement is based on a long history of controlling trade in endangered species through the issue of export permits (species listed in appendices 1 and 2). It adds the twist of requiring an import permit for an export permit to be issued, in order to prevent circumvention to non-parties.

Montreal Protocol on Substances that Deplete the Ozone Layer, 1987. The Montreal Protocol is an application agreement to the Vienna Convention regarding substances which cause depletion of the ozone layer (1985)

The parties have agreed to reduce CFC production by 50 per cent by 1999. During negotiations in London in 1990, requirements were further restricted and completely new substances were included in the agreement. It has been decided that CFC production will stop completely by 1996. The agreement will take effect three months after a sufficient number of countries have ratified it; in reality this takes approximately two years after an agreement is signed. In March 1993, 111 countries had acceded to the 87-year agreement. Among those that had not signed were South Korea, Colombia and Vietnam. Brazil signed in 1990, as did Chile and Argentina.

In 1992, 21 more countries signed, including India, Indonesia and Israel. The amendments proposed in London in 1990 were ratified by 51 countries, mostly the industrialized countries but also Chile, China, India and Mexico. No country has yet ratified the changes made in Copenhagen in November 1992. The parties agree, after a certain date, not to export or import specific substances to non-parties and to ban importation of CFC-containing products as of 1 January 1993.

Basel Convention on the Control of Transboundary Movements of Hazardous Wastes and their Disposal, 1989

Each party has the right to prohibit the import of hazardous wastes. Export should only be permitted when the importing country's government has given permission in writing. If there is reason to believe that the waste will not be disposed of in an 'environmentally sound manner', then it should not be exported. Trade with non-signatory countries is not allowed.

Sources:
International Trade 1990–1991, Appendix 1, Trade and the Environment, GATT 1992
Frihandeln ett hot mot miljöpolitiken eller tvärtom? Ds 1992:12
Status of Ratification of I. The Vienna Convention for. . ., II. The Montreal Protocol on . . ., III. The Amendment to the Montreal Protocol on . . ., UNEP 1993.

Appendix 2

EXCERPTS FROM THE RIO DECLARATION

Principle 2

States have, in accordance with the Charter of the United Nations and the principles of international law, the sovereign right to exploit their own resources pursuant to their own environmental and developmental policies, and the responsibility to ensure that activities within their jurisdiction or control do not cause damage to the environment of other states or of areas beyond the limits of national jurisdiction.

Principle 11

States shall enact effective environmental legislation. Environmental standards, management objectives and priorities should reflect the environmental and developmental context to which they apply. Standards applied by some countries may be inappropriate and of unwarranted economic and social cost to other countries, in particular developing countries.

Principle 12

States should cooperate to promote a supportive and open international economic system that would lead to economic growth and sustainable development in all countries, to better address the problems of environmental degradation. Trade policy measures for environmental purposes should not constitute a means of arbitrary or unjustifiable discrimination or a disguised restriction on international trade. Unilateral actions to deal with environmental challenges outside the jurisdiction of the importing country should be avoided. Environmental measures addressing transboundary or global environmental problems should, as far as possible, be based on an international consensus.

Principle 13

States shall develop national law regarding liability and compensation for the victims of pollution and other environmental damage. States shall also cooperate in an expeditious and more determined manner to develop further international law regarding liability and compensation for adverse effects of environmental damage caused by activities within their jurisdiction or control to areas beyond their jurisdiction.

Principle 16

National authorities should endeavour to promote the internalization of environmental costs and the use of economic instruments, taking into account the approach that the polluter should, in principle, bear the cost of pollution, with due regard to the public interest and without distorting international trade and investment.

Appendix 3

EXCERPTS FROM GATT

From Article 3: National Negotiations Regarding the Question of Internal Taxation and Regulation

1. The contracting parties recognize that internal taxes and other internal charges, and laws, regulations and requirements affecting the internal sale, offering for sale, purchase, transportation, distribution or use of products, internal quantitative regulations requiring the mixture, processing or use of products in specified amounts or proportions, should not be applied to imported or domestic products so as to afford protection to domestic production.
4. The products of the territory of any contracting party imported into the territory of any other contracting party shall be accorded treatment no less favourable than that accorded to *like products* of national origin in respect of all laws, regulations and requirements affecting their internal sale, offering for sale, purchase, transportation, distribution or use. The provisions of this paragraph shall not prevent the application of differential internal transportation charges which are based exclusively on the economic operation of the means of transport and not on the nationality of the product.

From Article 11: General Elimination of Quantitative Restrictions

1. No prohibitions or restrictions other than duties, taxes or other charges, whether made effective through quotas, import or export licenses or other measures, shall be instituted or maintained by any contracting party on the importation of any product of the territory of any other contracting party or on the exportation or sale for export of any product destined for the territory of any other contracting party.
2. The provisions of paragraph 1 of this Article shall not extend to the following:

(a) Export prohibitions or restrictions temporarily applied to prevent or relieve critical shortages of foodstuffs or other products essential to the exporting contracting party;

(b) Import and export prohibitions or restrictions necessary to the application of standards or regulations for the classification, grading or marketing of commodities in international trade.

Article 20: General Exceptions

Subject to the requirement that such measures are not applied in a manner which would constitute a means of arbitrary or unjustifiable discrimination between countries where the same conditions prevail, or a disguised restriction on international trade, nothing in this agreement shall be construed to prevent the adoption or enforcement by any contracting party of measures:

(a) necessary to protect public morals;
(b) necessary to protect human, animal or plant life or health;
(c) relating to the importation of gold or silver;
(d) necessary to secure compliance with laws or regulations which are not inconsistent with the provisions of this Agreement, including those relating to customs enforcement, the enforcement of monopolies operated under paragraph 4 of Article II and Article XVII, the protection of patents, trade marks and copyrights, and the prevention of deceptive practices;
(e) relating to the products of prison labour;
(f) imposed for the protection of national treasures of artistic, historic or archaelogical value;
(g) relating to the conservation of exhaustible natural resources if such measures are made effective in conjunction with restrictions on domestic production or consumption;
(h) undertaken in pursuance of obligations under any inter-governmental commodity agreement which conforms to criteria submitted to the CONTRACTING PARTIES and not disapproved by them or which is itself so submitted and not so disapproved;
(i) involving restriction on exports of domestic materials necessary to assure essential quantities of such materials to a domestic processing industry during periods when the domestic price of such materials is held below the world price as part of a governmental stabilization plan; *provided* that such restrictions shall not operate to increase the exports of or the protection afforded to such domestic industry, shall not depart from the provisions of this Agreement relating to non-discrimination;
(j) essential to the acquisition or distribution of products in general or local short supply; *provided* that any such measures shall be consistent with the principle that all contracting parties are entitled to an equitable share of the international supply of such products, and that any such measures, which are inconsistent with the other provisions of this Agreement shall be discontinued as soon as the conditions giving rise to them have ceased to exist. The CONTRACTING PARTIES shall review the need for the sub-paragraph not later than 30 June 1960.

Article 21: Security Exceptions

Nothing in this Agreement shall be construed:

(a) to require any contracting party to furnish any information the disclosure of which it considers contrary to its essential security interests; or

(b) to prevent any contracting party from taking any action which it considers necessary for the protection of its essential security interests

  (i) relating to fissionable materials or the materials from which they are derived;

  (ii) relating to the traffic in arms, ammunition and implements of war and to such traffic in other goods and materials as is carried on directly or indirectly for the purpose of supplying a military establishment;

  (iii) taken in time of war or other emergency in international relations; or

(c) to prevent any contracting party from taking any action in pursuance of its obligations under the United Nations Charter for the maintenance of international peace and security.

Appendix 4

EXCERPTS FROM THE TREATY OF ROME

Article 100a4

4. If, after the adoption of a harmonization measure by the Council acting by a qualified majority, a Member State deems it necessary to apply national provisions on grounds of major needs referred to in Article 36, or relating to protection of the environment or the working environment, it shall notify the Commission of these provisions.

The Commission shall confirm the provisions involved after having verified that they are not a means of arbitrary discrimination or a disguised restriction on trade between Member States.

By way of derogation from the procedure laid down in Articles 169 and 170, the Commission or any Member State may bring the matter directly before the Court of Justice if it considers that another Member State is making improper use of the powers provided for in this Article.

Article 130r

1. Action by the Community relating to the environment shall have the following objectives:

(i) to preserve, protect and improve the quality of the environment;
(ii) to contribute towards protecting human health;
(iii) to ensure a prudent and rational utilization of natural resources.

2. Action by the Community relating to the environment shall be based on the principles that preventive action should be taken, that environmental damage should as a priority be rectified at source, and that the polluter should pay. Environmental protection requirements shall be a component of the Community's other policies.

3. In preparing its action relating to the environment, the Community shall take account of:

(i) available scientific and technical data;
(ii) environmental conditions in the various regions of the Community;
(iii) the potential benefits and costs of action or lack of action;
(iv) the economic and social development of the Community as a whole and the balanced development of its regions.

4. The Community shall take action relating to the environment to the extent to which the objectives referred to in paragraph 1 can be attained better at Community level than at the level of the individual Member States. Without prejudice to certain measures of a Community nature, the Member States shall finance and implement the other measures.

5. Within their respective sphere of competence, the Community and the Member States shall cooperate with third countries and with the relevant international organizations. The arrangements for Community cooperation may be the subject of agreements between the Community and the third parties concerned, which shall be negotiated and concluded in accordance with Article 228.

The previous paragraph shall be without prejudice to Member States' competence to negotiate in international bodies and to conclude international agreements.

Article 130s

The Council, acting unanimously on a proposal from the Economic and Social Committee, shall decide what action is to be taken by the Community.

The Council shall, under the conditions laid down in the preceding subparagraph, define those matters on which decisions are to be taken by a qualified majority.

Article 130t

The protective measures adopted in common pursuant to Article 130s shall not prevent any Member State from maintaining or introducing more stringent protective measures compatible with this Treaty.

Appendix 5

TRADE AND ENVIRONMENT IN GATT

ANNEX II

Trade and Environment
Decision 14 April 1994

Ministers, meeting on the occasion of signing the final Act embodying the results of the Uruguay Round of Multilateral Trade Negotiations at Marrakesh on 15 April 1994,

Recalling the preamble of the Agreement establishing the World Trade Organisation (WTO), which states that members' 'relations in the field of trade and economic endeavour should be conducted with a view to raising standards of living, ensuring full employment and a large and steadily growing volume of real income and effective demand, and expanding the production of and trade in goods and services, while allowing for the optimal use of the world's resources in accordance with the objective of sustainable development, seeking both to protect and preserve the environment and to enhance the means for doing so in a manner consistent with their respective needs and concerns at different levels of economic development.'

Noting:

- the Rio Declaration on Environment and Development, Agenda 21, and its follow-up in GATT, as reflected in the statement of the Chairman of the Council of Representatives to the CONTRACTING PARTIES at their 48th Session in December 1992, as well as the work of the Group on Environmental Measures and International Trade, the Committee on Trade and Development, and the Council of Representatives;

- the work programme envisaged in the Decision on Trade in Services and the Environment; and
- the relevant provisions of the Agreement on Trade-Related Aspects of Intellectual Property Rights,

Considering that there should not be, nor need be, any policy contradiction between upholding and safeguarding an open, non-discriminatory and equitable multilateral trading system on the one hand, and acting for the protection of the environment, and the promotion of sustainable development on the other,

Desiring to coordinate the policies in the field of trade and environment, and this without exceeding the competence of the multilateral trade system, which is limited to trade policies and those trade-related aspects of the environmental policies which may result in significant trade effect for its members.

Decide:

- to direct the first meeting of the General Council of the WTO to establish a Committee on Trade and Environment open to all members of the WTO to report to the first biennial meeting of the Ministerial Conference after the entry into force of the WTO when the work and terms of reference of the Committee will be reviewed, in the light of recommendations of the Committee,
- that the [Trade Negotiating Committee] TNC Decision of 15 December 1993 which reads, in part, as follows:
  (a) to identify the relationship between trade and environmental measures, in order promote sustainable development;
  (b) to make appropriate recommendations on whether any modifications of the provisions of the multilateral trading system are required, compatible with the open, equitable and non-discriminatory nature of the systems as regards, in particular:
    — the need for rules to enhance positive interaction between trade and environmental measures, for the promotion of sustainable development, with special consideration to the needs of developing countries, in particular those of the least developed among them; and
    — the avoidance of protectionist trade measures, and the adherence to effective multilateral disciplines to ensure responsiveness of the multilateral trading system to environmental objectives set forth in Agenda 21 and Rio Declaration, in particular Principle 12; and
    — surveillance of trade measures used for environmental purposes, of trade-related aspects of environmental measures which have significant trade effects, and of effective implementation of the multilateral disciplines governing those measures;

constitute, along with the preambular language above, the terms of reference of the Committee on Trade and Environment,

- that, within these terms of reference, and with the aim of making international trade and environmental policies mutually supportive, the Committee will initially address the following matters, in relation to which any relevant issue may be raised:
  - — the relationship between the provisions of the multilateral trading system and trade measures for environmental purposes, including those pursuant to multilateral environmental agreements;
  - — the relationship between environmental policies relevant to trade and environmental measures with significant trade effects and the provisions of the multilateral trading system;
  - — the relationship between the provisions of the multilateral trading system and:
    - (a) charges and taxes for environmental purposes
    - (b) requirements for environmental purposes relating to products, including standards and technical regulations, packaging, labelling and recycling;
  - — the provisions of a multilateral trading system with respect to the transparency of trade measures used for environmental purposes and environmental measures and requirements which have significant trade effects;
  - — the relationship between the dispute settlement mechanisms in the multilateral trading system and those found in multilateral environmental agreements;
  - — the effect of environmental measures on market access, especially in relation to developing countries, in particular to the least developed among them, and environmental benefits of removing trade restrictions and distortions;
  - — the issue of exports of domestically prohibited goods,
- that the Committee on Trade and Environment will consider the work programme envisaged in the Decision on Trade in Services and Environment and the relevant provisions of the Agreement on Trade-Related Aspects of Intellectual Property Rights as an integral part of its work, within the above terms of reference,
- that, pending the first meeting of the General Council of the WTO, the work of the Committee on Trade and Environment should be carried out by a Sub-Committee of the Preparatory Committee of the World Trade Organization (PCWTO), open to all members of the PCWTO,
- to invite the Sub-Committee of the Preparatory Committee, and the Committee on Trade and Environment when it is established, to provide input to the relevant bodies in respect of appropriate arrangements for relations with inter-governmental and non-governmental organizations referred to in Article V of the WTO.

Index